Elvis Pantaleão Ferreira
Fernando Cartaxo Rolim Neto
Luiza Teixeira de Lima Brito

Cistern water management for food production

A proposal in an area of the Brazilian semi-arid region

ScienciaScripts

Imprint

Cover image: www.ingimage.com

This book is a translation from the original published under ISBN 978-3-330-76202-2.

Publisher:
Sciencia Scripts
is a trademark of
Dodo Books Indian Ocean Ltd. and OmniScriptum S.R.L publishing group

120 High Road, East Finchley, London, N2 9ED, United Kingdom
Str. Armeneasca 28/1, office 1, Chisinau MD-2012, Republic of Moldova, Europe
Managing Directors: Ieva Konstantinova, Victoria Ursu
info@omniscriptum.com

Printed at: see last page
ISBN: 978-620-8-38326-8

Elvis Pantaleão Ferreira
Fernando Cartaxo Rolim Neto
Luiza Teixeira de Lima Brito

Cistern water management for food production

CONTENTS

CHAPTER 1

INTRODUCTION

Brazil's semi-arid region occupies 67% of the Northeast, totalling 1.133 municipalities in all states, as well as the north of the state of Minas Gerais. It is characterised by the lowest rainfall in the country, with average annual rainfall equal to or less than 800 mm, marked by spatio-temporal variability, low relative humidity of around 50%, vegetation basically made up of hyperxerophilous caatinga with stretches of deciduous forest and an average annual temperature of 26.3 °C, ranging from 18.2 to 34 °C (BRASIL, 2005; Correia et al., 2011).

The low availability of water in the region has been an obstacle to families staying in rural areas. However, it is recognised that this region has potential that can be better exploited for the sustainable development of its populations (PAN BRASIL, 2004). In this sense, research shows that in order to solve the water problem in the Brazilian semi-arid region, rainwater should be considered as important as or even more important than other available water sources. This can be done through the development of research and the use of technologies to increase the supply of water in the region in order to adequately meet the different demands.

Over the years, various social technologies have been developed, consolidated and improved to strengthen the coexistence of families in semi-arid conditions, based on academic knowledge and founded on the valuable contribution of the accumulated knowledge and experiences of the people who live there, given their creativity in finding practical solutions to everyday problems. According to Costa (2013), social technologies aimed at the semi-arid region can be defined as simple, low-cost alternatives that are easy to reapply to solve structural and effective social problems. In this context, Brava (2004) and EL-Deir (2013) emphasise that social technologies are basically based on two premises for their propagation: the effective participation of the community in their construction and/or appropriation process and the sustainability of the solutions presented.

Among the technologies that can be used is the "calçadão" cistern, which is part of the "Uma Terra e Duas Águas" (One Land and Two Waters) Programme - P1+2. In this programme, the cisterns are built in a semi-underground cylindrical shape, in pre-moulded cement slabs, with a capacity of 52,000 litres, supplied by rainwater captured by a 200 m boardwalk[2] , built in masonry. They are also called production cisterns, as they are intended for productive use and are installed in the backyards of rural homes.

Few studies have been carried out and published on the management of rainwater stored in P1+2 cisterns for food production in the family farming system in the Brazilian semi-arid region. There is therefore a need for research that contributes to the production of fruit and vegetables for family consumption, grown using cistern water in the semi-arid region of Pernambuco. Almost all of the studies on the efficient use of water do not include the use of cisterns as a field of action, as they are aimed at irrigated fruit growing in the São Francisco Valley, which is highly productive, motivated by generous investments and the need to increase productivity and fruit quality in the context of a globalised economy.

The aim of this study was to evaluate the production of fruit and vegetable crops grown with water from the P1+2 cistern in the semi-arid region of Pernambuco, which are exploited in the family farming system. To this end, the research consisted of species commonly grown in rural areas by families in the region. At the same time, the aim was to find out the current situation regarding the use of P1+2 cisterns in the rural communities covered by the programme, located in the microregion of the Vale do Submédio do São Francisco, and their real contribution to improving the quality of life of families, with a view to providing subsidies to help strengthen the Uma Terra e Duas Águas Programme in the region.

CHAPTER 2

THE P1+2 PROGRAMME

The semi-arid northeast of Brazil, like any other semi-arid region in the world, will always be subject to periodic droughts. This is because one of the natural characteristics of this type of climate is the occurrence of irregular and poorly distributed rainfall. Thus, for the sertanejo to live in these conditions, it is necessary to adopt technologies for coexistence with the region, such as capturing rainwater to use it both during the veranicos that occur during the rainy season and afterwards (Moura et al., 2007). However, it is recognised that this region has potential that can be better exploited for the sustainable development of its populations (PAN BRASIL, 2004).

Various social technologies have been developed, consolidated and improved to strengthen coexistence with the semi-arid region, based on academic knowledge and grounded in the valuable contribution of knowledge from the accumulated experiences of the people who live there, given their creativity in finding practical solutions to the daily problems of living in the semi-arid region.

Social technology includes the introduction of reapplicable techniques or methodologies, developed in interaction with the community and which represent effective solutions for social transformation, which can combine popular knowledge, social organisation and technical-scientific knowledge. It is based on the dissemination of low-cost, easy-to-apply solutions to everyday problems such as food, income, health and water resources, among others, taking into account collective participation in the organisation, development and implementation process (FBB, 2015).

Thus, several social technologies of proven efficiency that adopt strategies for living in the semi-arid region are being reapplied by Non-Governmental Organisations (NGOs) linked to the Brazilian Semi-Arid Articulation (ASA), including the 52,000-litre plate cisterns of the One Land, Two Waters Programme (P1+2).

The 52,000-litre slab cistern is also known as the "pavement cistern" of the P1+2.

These are semi-underground cylindrical tanks made of pre-moulded cement slabs (Diaconia, 2008). They are called production cisterns because they are designed to store rainwater for productive use, and are commonly installed in the backyards of rural homes. In this programme, the water reservoir is connected to a 200 m long boardwalk[2] , built of masonry, which serves as a catchment area for rainwater (Figure 1).

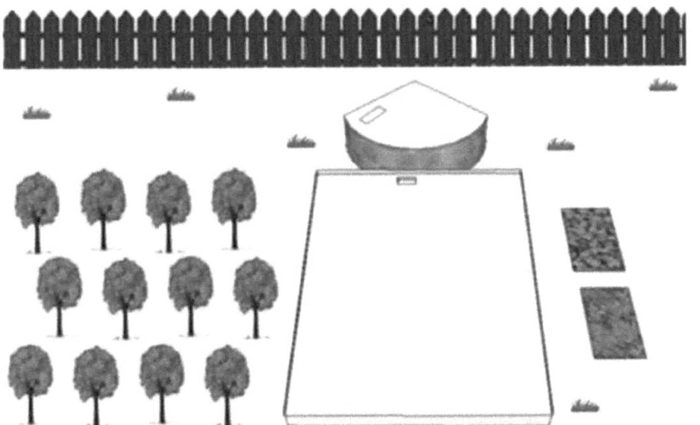

Figure 1 - Schematic of a P1+2 production cistern

Prepared by Elvis Pantaleão Ferreira (2015).

In the One Land, Two Waters Programme (P1+2), 1 stands for land for production and 2 for two types of water - drinking water for human consumption and water for food production. The families to benefit must have access to water for human consumption, like the cisterns in the One Million Cisterns Programme (P1MC), and must also meet the following criteria: women must be heads of household; families with children aged 0 to 6; children and adolescents attending school; adults aged 65 or over, or a family member living in the household with special needs (ASA, 2014).

The following requirements must also be taken into account when choosing the families and the type of technology best suited to their reality: the characteristics of the soils; the geology through the formation of crystalline rock, sediment or sandstone; the location of the implementations; the logic of production, be it agriculture, livestock or extractivism; and the forms of management (ASA, 2014).

The P1+2 is a training and social mobilisation project for coexistence with the Brazilian semi-arid region, whose strategic objective is to ensure that the rural population has access to land and water, both for family and animal consumption and for food production, promoting food, animal and plant safety by building participatory processes for the rural population (Gnadlinger et al., 2007).

The benchmark for PI+2 is the "1-2-1 Programme" developed in China from the 1990s onwards in the semi-arid region of Gansu State, which made it possible to use rainwater stored in reservoirs to save plants in times of rain scarcity, making it possible to grow vegetables and fruit in perpetuity and boosting small animal husbandry. Given the similarity of the environmental conditions to the Brazilian semi-arid region, this experience was presented at the end of the 1990s in Brazil, at the Brazilian Symposium on Rainwater Harvesting and Management, in Petrolina - PE (Brito et al., 2010).

This technology has been experimented with by society in the Brazilian semi-arid region and embraced as a public policy, which has received special attention from the federal government, research institutions and non-governmental organisations, with the aim of guaranteeing a sustainable increase in the supply of water for growing food, water and food security for the population of the Brazilian semi-arid region, providing opportunities to improve quality of life and keeping the rural population in the countryside.

Thus, the advancement of technological knowledge for living in the semi-arid region, incorporated into the body of knowledge accumulated by the region's inhabitants, has made it possible to bring low-cost, proven social technologies to the countryside, enabling greater access to water and progress in the productive structuring of the property.

By December 2014, 3,080 52,000-litre cisterns had been built in the semi-arid region. The municipality of Juazeiro, in Bahia, has the largest number of cisterns, with 469 units having been built, followed by the municipality of São José do Egito, in

Pernambuco, with 184 units (ASA, 2014). This scenario represents an important step forward in articulating actions and implementing solutions capable of expanding access to water and contributing to the region's socio-economic development.

The Ministry of Social Development (MDS), together with the Ministry of National Integration (MI) and other federal executive bodies, have been maximising and prioritising the application of federal public policy resources to the semi-arid region. Various civil society organisations linked to the ASA and research centres have also joined forces to restore self-esteem and citizenship in this area (MI, 2009).

CHAPTER 3

MATERIAL AND METHODS

3.1 Characterisation of the study area

The area is located in the semi-arid region of the state of Pernambuco, specifically in the micro-region of the Vale do Submédio São Francisco, in the municipality of Petrolina. It is bordered to the north by Dormentes, to the south by the state of Bahia, to the east by Lagoa Grande, and to the west by the state of Bahia and Afrânio (Figure 2). The seat of the municipality is geographically located at 09° 24' 04" south latitude and 40° 30' 16" west longitude, about 760 km from Recife, the state capital, which can be reached via BRs 122/428.

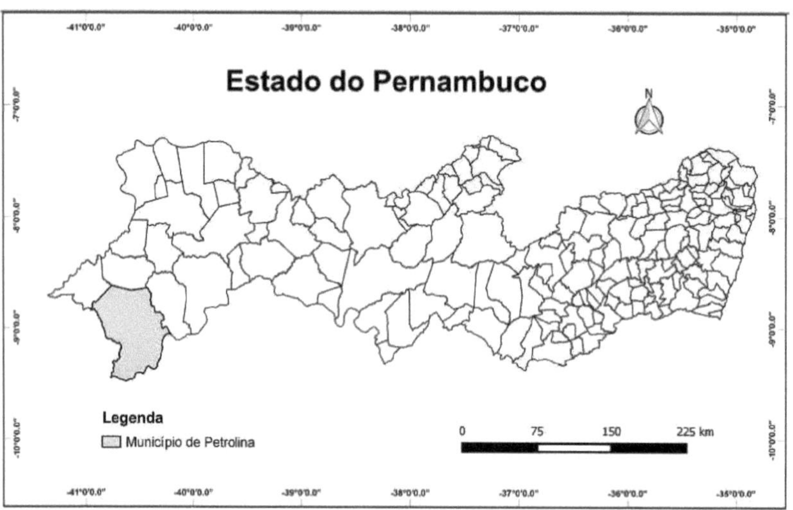

Figure 2 - Spatial location of the municipality of Petrolina in the state of Pernambuco

Prepared by Elvis Pantaleão Ferreira (2015).

The study area is specifically located in the Caatinga Experimental Field, belonging to the Brazilian Agricultural Research Corporation - Embrapa Semiárido, geographically located at 09° 43' 03" S and 40° 19' 37" W, at an altitude of 383 metres, with a landscape characteristic of the northeastern semi-arid region. According to the

Koppen classification system, the region's climate is of the BSwh type - semi-arid climate, characterised by being dry and very hot, with average annual rainfall of 549 mm, with around 90% of the rainfall concentrated between November and April. The average annual temperature is 26.3 °C, ranging from 18.2 to 34 °C (Teixeira, 2010).

3.2Experiment , soil, cultivation and water

The experiment was conducted in the field and consisted of five species of fruit trees commonly grown by rural families, namely mango (Mangifera indica L.), pine (Annona squamosa), cashew (Anacardium occidentale), acerola (Malphigia glabra L.) and orange (Citrus sinensis), four years old, at a spacing of 5 metres between plants and 5 metres between rows. The crop design in the field consisted of two treatments, irrigated[1] and non-irrigated, with three replications for the irrigated treatment and two replications for the non-irrigated treatment, with one plant of each species corresponding to a repetition, according to the [(3x5) + (2x5)] arrangement, totalling 25 plants (Figure 3).

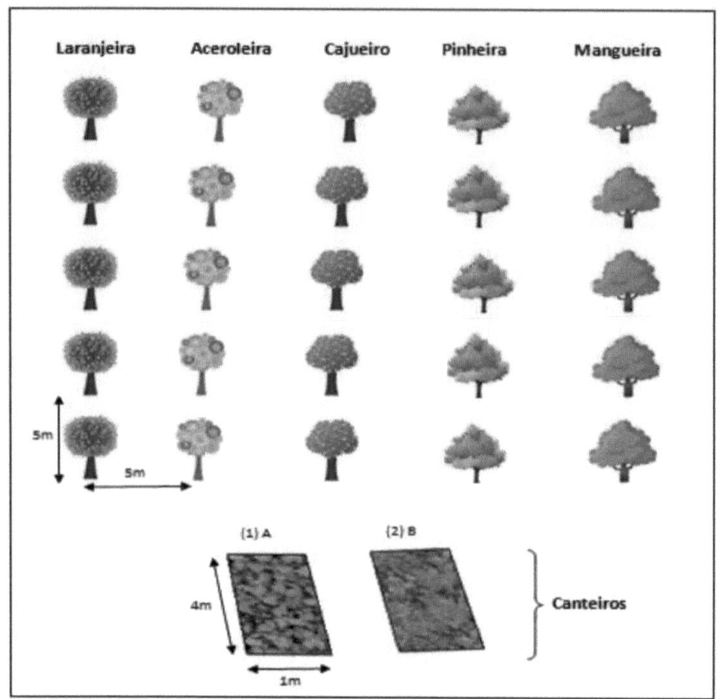

Figure 3 - Field experiment layout

Prepared by Elvis Pantaleão Ferreira (2015).

The experiment also included the evaluation of two beds (1) A and (2) B, of the leirão type, of four square metres - 4 m[1] [2] each, cultivated with vegetables in order to also evaluate the possibility of cultivating olive species under different water levels, commonly grown by rural families in the region, such as peppers (Capsicum annuum), cabbage (Brassica oleracea), rocket (Eruca sativa), coriander (Coriandrum sativum) and lettuce (Lactuca sativa), as can also be seen in Figure 3.

The beds were cultivated in alternating cycles, with the vegetables mentioned

[1] The term "irrigated" used in this work refers to the application of a minimum volume of water in due to the limited water storage capacity of the P1+2 cistern. Thus, this volume does not correspond to the evapotranspirometric demand of the plants, but only to keep them alive and produce to meet the families' consumption.

above in each sowing cycle until the plants germinated. The beds were covered with palm straw to minimise evaporation and prevent the seeds from being eaten by birds. The areas of the beds were covered with black sombrite, 60 per cent shading, in order to reduce insolation and slow down the flow of evapotranspiration. This is a common practice among family farmers in the region who, when they don't have shading, use other covering materials, such as coconut leaves and palm trees, as observed during field visits.

The soil in the experiment area was classified as Argissolo Amarelo Eutrófico abrúptico plíntico (Santos et al., 2006).

As for crop management, the studies were conducted in such a way as to represent as closely as possible the reality of family farmers in the semi-arid region. Therefore, weeding and manual mowing of spontaneous vegetation was carried out in the experiment area when necessary. At the start of the rainy season, only organic fertiliser was applied, consisting of 10 kg of tanned goat manure per plant and 250 kg of NPK in the 10-10-12 formulation. In the area of the beds, 4 $L.m^{-2}$ of tanned goat manure was applied, mixed well with the soil. In the orchard, mulch was applied to the crown of each plant, using coconut bagasse, with the aim of reducing water loss through evaporation and the incidence of spontaneous plants that also compete with the fruit trees for water and nutrients in the soil, which are considered to be limiting factors.

The water used to carry out this work came from rainfall in 2013, intercepted by a rainwater catchment area made up of a 200 metre boardwalk[2] , built from masonry. This infrastructure makes up the production cistern, which has a capacity of 52,000 litres. Once the cistern was completely full, it did not receive any more water during the research.

The research into water management applied to fruit and vegetable crops was conducted in the field from January to December 2014.

3.4 Applying water to fruit and vegetable plants

To define the amount of water applied to the fruit trees and vegetables, the main criterion was to limit the volume of the cistern to 52,000 litres. Based on this, the volumes to be applied to the plants were dimensioned, with 15 fruit trees of different species and the area exploited by the beds at 8 m^2 , considered sufficient to produce fruit and vegetables to meet the needs of rural families.

These considerations were based on previous field studies carried out by Brito et al. (2012) and Cavalcanti et al. (2012), as well as research carried out at Embrapa Semiárido, where it is possible to obtain yields throughout the year, even in drought situations such as those that occurred in 2012/2013. Therefore, the aim was to establish an increasing volume applied, so that the crops would not lack water throughout the year, depending on the rainfall.

To this end, as discussed by Brito et al. (2010), three periods of the year were established: the Rainy Period (PC), the Intermediate Period (PI) and the Dry Period (PS), between the months of January to April, May to August and September to December, respectively, for the conditions of the municipality of Petrolina - PE.

In the Rainy Period - PC the volume of water applied to the orchard was 8 litres per plant, three times a week, for 14 weeks of the year; in the Intermediate Period - PI each plant received 12 litres of water, applied at the same frequency for 18 weeks, and in the Dry Period 16 litres of water were applied per plant, three times a week for 20 weeks, as shown in Table 1.

Table 1 - Estimated cistern water volume applied to the 15 fruit trees

Time of year	No. of plants	No. of Weeks	Frequency of application/week	Vol. of water applied (L)	Total volume applied (L)
PC	15	14	3	8	5.040
PI	15	18	3	12	9.720
PS	15	20	3	16	14.400
Total volume of water applied to the orchard					29.160

PC = rainy season; PI = intermediate season; PS = dry season.

With regard to the volume of water applied to the vegetables grown in beds A and B, the strategy used for bed "A" was to apply 32 litres of water per day, half in the morning and half in the afternoon. In bed "B", 50% of this volume of water was applied, following the same criteria used in the first bed (Table 2), which corresponds to a 4 mm day^{-1} , given that the beds have an area of 4 m^2 each.

Table 2 - Estimated volume of cistern water applied to vegetables

Days of the year	Site area (m)^2	Blade applied (mm **day-1**)	% of Volume applied	Volume applied (L)
365	4	8	100%	11.680
365	4	4	50%	5.840
Total volume of water applied		in vegetable beds		17.520

Water was applied to the orchard by gravity, after the water from the cistern was fed into a 2,000 litre water tank situated on concrete pillars about two metres above the ground, where it was connected to a 1 inch line filter, followed by a 25 mm main pipe. Perpendicular to the fruit trees, 12.7 mm polyethylene hoses were laid down with 8.0 mm catifa drippers, with a plant spacing of 5.0 metres.

For all the fruit trees in the orchard, the volumes applied were dripped every other day (Figure 5A). To calculate the volume applied, plastic containers were used as a strategy, measured with the volume to be applied in each period - 8, 12 and 16 litres. When the reference level was reached, water distribution was stopped (Figure 5B). This strategy saves the grower worrying about the pre-established water application time. For the vegetable beds, water was applied manually, using a plastic watering can. In both the orchard and the vegetable beds, when there was a volume of rainfall greater than or equal to the volume to be irrigated in any period, irrigation was suspended for that day. The total amount of water used was 46,680 litres. There was therefore a balance of 5,320 litres of water inside the cistern, which is important to compensate for losses due to evaporation and infiltration, and to conserve the structure's humidity, preventing physical damage such as the appearance of microcracks. This total volume applied was in addition to the rainwater.

Figure 4 - Method of applying water to the fruit trees (A); technique used to control the volume of water applied - intermediate period (B)

Photo: Elvis Pantaleão Ferreira (2014).

3. 3 Determination of the soil's physical attributes

Soil samples were collected from the experiment area in the 0.0 to 0.2 m and 0.2 to 0.4 m layers in order to carry out physical and chemical characterisation. The latter layer covers the effective depth of the root system for fruit crops (Pires et al., 1999). The deformed samples were collected using an auger, placed in plastic bags and duly identified. Samples with preserved structure were collected using an Uhland-type auger made up of stainless steel cylinders with one of the cutting edges measuring 50 mm in diameter by 53 mm in height, which, after collection, were duly identified and packaged (Figure 5).

Figure 5 - Equipment used for soil collection

Photo: Elvis Pantaleão Ferreira (2014).

a) **Particle size** - This was determined using the pipette method according to the recommendations of Embrapa (1997), by dispersing 20 g of Fine Air-Dried Soil (TFSA) with 1 mol L sodium hydroxide .$^{-1}$

The sands were separated into coarse, medium and fine sand by wet sieving on sieves with mesh sizes of 1.00 - 0.50 mm; 0.50 - 0.25 mm; and 0.25 - 0.053 mm in diameter and fractionated according to the granulometric classification adopted by the Brazilian Society of Soil Science - SBSC (Amaral Filho et al,

2008). Based on Stokes' Law, the sedimentation time for the silt and clay particles was calculated.

b) **Water-dispersed clay** - Same method as for the particle size analysis, with only the chemical dispersant removed (Embrapa, 1997).

c) **Degree of Flocculation** - Obtained using the equation below (Embrapa, 1997).

$$GF = \frac{\textit{(total clay-water dispersed clay)}}{\textit{total clay}} * 100 \qquad \text{Equation (1)}$$

d) **Soil Porosity** - Obtained using the equation below (Embrapa, 1997).

$$p = \left(1 - \frac{Ps}{Pp}\right) \qquad \text{Equação (2)}$$

Where p = porosity ($cm^3 .cm^{-3}$); Pp = particle density ($g.cm^{-3}$); Ps = soil density

($g.cm^{-3}$).

e) Soil Density - This was obtained **using the** conventional "volumetric ring" method, based on the equation below (Embrapa, 1997).

$$Ds = \frac{ms}{Vc}\ ;\ onde\ Ds = \frac{ms}{Ab\ x\ h} \qquad \text{Equação (3)}$$

Where, Ds = soil density in $g.cm^{-3}$; ms: dry soil mass in (g) after drying at 105 - 110 °C; Vs = cylinder volume in cm^{-3} ; Ab = cylinder base area in cm^2 ; h = cylinder height (cm).

f) Particle Density - Determined using the volumetric flask method (Equation 4). based on the equation below (Embrapa, 1997).

$$Dp = \frac{Ms}{50 - Va} \qquad \text{Equação (4)}$$

Where, Dp = density of solids in $g.cm^{-3}$; Ms = mass of soil in g; and Va = volume spent on the sample in cm .^{3}

g) Determining the Electrical Conductivity of the Soil - The electrical conductivity of the soil's saturation extract (ECes) was determined using a 1:2.5 soil/water extract ratio. To do this, 20 g of air-dried fine soil was weighed into suitable containers and 50 ml of distilled water was added. They were then shaken in a Wagner shaker at 50 rpm for 15 minutes. After this procedure, the samples were left to rest for 60 minutes, and only then were stirred again for the same length of time. After the second stirring, electrical conductivity readings were taken using a previously calibrated digital bench conductivity meter (MOLIN et al., 2005).

h) Soil Water Characteristic Curve - Six samples of soil with an undeformed structure were collected from the 0.2 m and 0.4 m layers and used to obtain the respective humidities at tensions of 1, 3, 6, 10, 333 and 1500 kPa. The equipment used in the laboratory was a tension table for the low tension points, followed by the

Richards pressure chamber (Klute, 1986).

After equilibration, the samples were removed from the equipment and their masses determined by weighing. At the end, the samples were dried in an oven at 105 - 110° C for 48 hours, and after thermal equilibrium in a desiccator, the masses of solids were determined. The soil density and moisture content were then calculated on a mass and volume basis. In this way, data was obtained for constructing the characteristic soil water retention curve.

The volume-based moisture data (6) was tabulated and adjusted as a function of the water tension in the soil, using the potential model proposed by Van Genuchten (1980), using the function that relates the matric potential to the soil moisture content (Equation 5). The **Retention Curve** - RetC software was used to determine the parameters of this equation.

$$\theta = \theta r + \frac{(\theta s - \theta r)}{[1 + (\alpha . \Psi)^n]^m} \qquad \text{Equação (5)}$$

Where: θ refers to the current volumetric water content, in $cm^3 .in^{-3}$; θr is the residual moisture in $cm^3\ cm^{-3}$; θs is the saturation moisture, in $cm^3\ cm^{-3}$; ψ is the tension with which water is retained in the soil in kPa; α, m and n are empirical constants. To calculate the value of m, the expression m = 1 - 1/n was used, as recommended by Mualem (1976).

3.5 Determination of soil chemical attributes

The methods used to determine the chemical analysis of the soil, described below, followed the Manual of Chemical Analysis of Soils, Plants and Fertilisers, published by Embrapa Solos (Embrapa, 1999). Thus, after collection, the composite samples were sent to the soil laboratory, air-dried, crumbled, passed through a sieve with a 2 mm mesh, thus obtaining the Fine Air-Dried Soil (TFSA) and the chemical analyses of pH in CaCl2, exchangeable cations (Ca^{2+} , Mg^{2+} , K^+ and Na^+), exchangeable acidity (Al^{3+}), potential acidity (H^+ ; Al^{3+}), organic matter content and available phosphorus.

The hydrogen potential pH was determined using the potentiometric principle of the effective concentration of H ions^{+} in the soil solution, using a combined electrode immersed in a suspension of soil/calcium chloride CaCl2 0.01 mol L^{-1} in a 1:2.5 ratio. The aim was to reduce the possible seasonal effects of salts that interfere with soil pH measurements due to the increase in electrolyte concentration (Ebeling et al., 2008).

The exchangeable cations Al^{3+} , Ca^{2+} , Mg^{2+} were determined using potassium chloride solution KCl 1 mol L^{-1} , with Al^{3+} determined volumetrically using sodium hydroxide solution NaOH 0.025 mol L^{-1} and bromothymol blue solution as an indicator.

Ca^{2+} and Mg^{2+} were determined using the complexometric method with a solution of ethylenediamine tetraacetic acid - EDTA 0.025 mol L^{-1} , Eriochrome-T black and carbonic acid as an indicator. Na^{+} and K^{+} were determined using the Mehlich-1 double acid extraction solution, which consists of a mixture of acids (HCl 0.05 mol L^{-1} + H2SO4 0.0125 mol L^{-1}) and determined using the direct flame photometry method.

The potential acidity of the soil H $+Al^{+3+}$ was extracted using a solution of calcium acetate [(CH3COO)2.Ca.H2O)] 0.5 mol L^{-1} at pH 7.0 and determined by alkalimetric titration of the extract with a solution of sodium hydroxide NaOH 0.025 mol L^{-1} , using phenolphthalein as an indicator.

Available phosphorus in the soil was determined by colourimetry using the phospho-molybdic complex method, using ascorbic acid (vitamin C) as a reductant, with Mehlich-1 as the extractant. The addition of a molybdate solution in an acidic medium to a solution containing phosphorus leads to the formation of a colourless phospho-molybdic complex which, after reduction, turns blue. The intensity of the blue colour is a function of the concentration of P in the solution.

Organic matter was determined using the Walkley-Black method by oxidising CO_2 using a solution of potassium dichromate ($K_2Cr_2O_7$) 0.2 mol L^{-1} and concentrated sulphuric acid (H_2SO_4), heated in a digester block and determined by titrating the amount of dichromate that had not yet been oxidised using ammoniacal ferrous sulphate "*Mohr's salt*" Fe $(NH_4)_2(SO_4)_2.6H_2O$ at 0.05 mol L^{-1} in an acidic medium, using diphenylamine as an indicator.

The following values were obtained from the chemical analysis data: SB (Base Sum) = $K^+ + Mg^{+2} + Ca^{+2} + Na^+$; CTCefet. (Effective Cation Exchange Capacity) = SB + Al^{+3} ; CTCpot. (Potential Cation Exchange Capacity) = SB + (H + Al); V (Base Saturation) = SB / CTCefet.; m (Aluminium Saturation) = (Al^{+3} / CTCefet.) x 100; C.O = M.O (Organic Matter) = C. O. x 1.724.

3.6 Determination of the physico-chemical attributes of the water

a) Electrical conductivity - In order to classify the water intended for irrigation with regard to the threat of soil salinisation, the electrical conductivity - EC - was determined. To do this, a sample of water was taken from inside the cistern and stored in a new 1000ml PET plastic bottle. In the laboratory, after manually stirring the water, an aliquot of 200 ml was taken to determine the electrical conductivity using a previously calibrated digital bench conductivity meter (Rheinheimer & Souza, 2000).

The analysis of the water's Sodium Absorption Ratio (SAR), which indicates the effect of exchangeable sodium on the soil's physical conditions in terms of the danger of alkalisation or sodification, was carried out at Embrapa Semiárido's agri-environmental laboratory.

b) Hydrogen Potential (pH) - After collecting a sample of water from inside the cistern and storing it in a new 1000 ml polyethylene terephthalate (PET) plastic bottle, it was determined using a bench pH meter, previously calibrated with standard solutions of 4.0, 7.0 and 10.0. In the laboratory, an aliquot of 200 ml was taken for pH determination (Silva et al., 2011). The reference value used in this study to classify the water was in accordance with the guidelines proposed for interpreting water quality for irrigation according to Ayers & Westcot (1999).

3.7 Monitoring soil water content

Volumetric soil moisture was measured indirectly using Frequency Domain Reflectometry (FDR*), a* non-destructive technique that allows repeatability in space and time (Marinho & Morais, 2009). According to Silva et al. (2008), this method is one of the most widely used for measuring soil moisture, as it provides reliable readings, quick responses and can take continuous measurements over long periods.

The volumetric soil moisture meter used was the *HH2* model, a *Delta - T* moisture profile sensor comprising a display with keyboard and data collector. According to the manufacturer's manual, it reads the volumetric soil moisture content in cm^3 cm^{-3} ; measuring scale 0.005 to 0.6 cm^3 cm^{-3} ; accuracy ± 0.005 cm^3 cm^{-3} , 0 to 70 °C, stabilisation time 1 second; response time < 0.5 seconds (Figure 6).

The soil moisture profile sensor has a sealed rod 0.035 m in diameter by 1.0 m long, with electronic sensors in the form of stainless steel rings, arranged and fixed at intervals of 0.10; 0.20; 0.30; 0.40; 0.60 and 1.0 m (Figure 6).

Figure 6 - Soil moisture content meter used

Photo: Elvis Pantaleão Ferreira (2014).

To measure soil moisture, the stem was inserted into an access tube, which was placed in the soil at a distance of 0.4 metres from the fruit tree stems (Figure 7), according to the recommendations of Nascimento et al. (2012). For the vegetables, the access tube was placed in the centre of the beds. When the humidity measurements were not being taken on site, the access tubes remained sealed to prevent evaporative loss or the entry of water and/or impurities.

Figure 7 - Visualisation of the access tube in the orchard for measuring soil moisture

Photo: Nilton de Brito Cavalcanti (2014); Elvis Pantaleão Ferreira (2014).

After inserting the probe into the access tube, a 100 MHz DC voltage analogue signal was applied to the steel rings, generating an electromagnetic field that extends approximately 0.1 m into the soil. The water content in the soil in the vicinity of the rings determines their dielectric properties, which are then converted into soil moisture content using the principle of reflectometry (DELTA - T, 2008).

Calibration data was assumed, according to studies carried out "in loco" by Nascimento et al. (2012). The water content in the soil was monitored throughout 2014, in all the species grown in the orchard, at intervals of 0.10, 0.20, 0.30, 0.40, 0.60 and 1.0 metres deep in each plant. The access tube was installed in the centre of the beds. The frequency of soil water content monitoring in the orchard took place in the morning, just before the water was applied, three times a week, at six depth intervals in the soil profile, on 25 plants over one year (52 weeks), according to the diagram [25x52x3x6], totalling 23,400 readings.

In both beds, soil water content was also monitored throughout 2014 in the morning, just before the water was applied, for 365 days, at six depth intervals, in two vegetable beds [365x6x2], totalling 4,380 readings.

Considering that the scenario presents a high water demand, the lower limit of soil water availability - factor (f) of 0.7 was discussed for the fruit group and 0.5 for vegetables, as discussed by Bernardo et al. (2013). The soil water availability factor is a parameter between field capacity and critical moisture for a given crop, which limits the amount of available soil water that the plant can utilise without causing major damage to productivity (Mantovani et al., 2013).

3.8 Analysing the data

The statistical treatment used to analyse the data was initially tabulated in an electronic spreadsheet and then submitted to Classical Descriptive Statistical Analysis - EDC, with the aim of observing the general behaviour of the data, using the calculation of statistical measures of central tendency of the universe surveyed. Exploratory Data Analysis (EDA) was then used to remove inconsistent *"outlierf" data, which* was eliminated based on the criteria of Vuolo (1996) and Cunha et al. (2002), who consider *"qutliers"* to be data below the lower barrier, estimated by: mean (X) minus two standard deviations (SD); and above the upper barrier: mean (X) plus two standard deviations (SD), according to the expression [(X - 2*SD);(X + 2*SD)]. Graphs were then drawn in order to infer the behaviour of the average volumetric humidity in the soil profile during the rainy, intermediate and dry periods.

3.9 Field survey of P1+2 communities

In conjunction with the research into rainwater management carried out in the field, a survey of the current situation regarding the use of production cisterns was carried out in three rural communities between June and October 2014: Lindolpho Silva Settlement (09° 04' 57"S and 40° 41' 28"W), Campo Verde (08° 34' 00" S and 41° 08' 20" W) and Baixa da Boa Vista (08° 30' 06"S and 40° 35' 12"W), located, respectively, in the municipalities of Petrolina, Afrânio and Dormentes, located in the microregion of the Vale do Submédio São Francisco (Figure 8), in the state of Pernambuco, which received P1+2 cisterns.

Figure 8 - Map of the state of Pernambuco with its micro-regions

Prepared by Elvis Pantaleão Ferreira (2015).

The evaluations consisted of finding out about routine activities relating to the use of water from the cistern in terms of the form, frequency and volume applied to fruit and vegetable crops, the equipment used, the production and management of the crops grown and their real contribution to improving the families' quality of life. The aim is also to provide information that will help strengthen the programme in the region and improve the quality of life of the families involved.

As methodological support, exploratory and qualitative research was carried out "in loco", where situations were observed and recorded as they occurred, through direct and interactive contact between the researcher and the people.

The main characteristic of the study was its descriptive approach. In addition, the semi-structured interview method was used, as recommended by May (2004) and Gil (2008). The interviews were guided by a pre-established information script and conducted using simple language suitable for the audience, in order to obtain detailed information that would later be used in a qualitative analysis. In these interviews, the researcher acted as a listener in a friendly and stimulating environment, allowing the interviewees to talk freely about the subject, generating information rich in accounts of their experiences, aspirations and opinions.

The first community visited was the Lindolpho Silva Settlement, located in Petrolina - PE, where the information gathered was obtained from the president of the Residents' Association, as well as complementary data obtained from dialogues with other people in the community. In the municipalities of Dormentes and Afrânio, the same methodology was used and interviews were carried out in the Baixa da Boa Vista and Campo Verde communities, respectively.

In 2010, the three communities involved in the research were given production cisterns of the "calçadão" type, built through agreements signed between the MDS and the ASA, as part of the federal government's One Land, Two Waters Programme (P1+2).

CHAPTER 4

RESULTS AND DISCUSSION

4.1 Rainfall and water quality

A survey of rainfall from 1963 to 2009 in the Petrolina region, published by Teixeira (2010), shows an average annual total of 549 mm. Graph 1 shows an overview of recent rainfall for the Petrolina - PE region between 2006 and 2014. During this period, the total annual average was 445.4 mm.

It is worth mentioning that in 2006, Embrapa Semiárido began pioneering research into the optimised use of rainwater for food production, using the production cistern as a social technology for living in the semi-arid region, similar to the P1+2 model used in federal government policy.

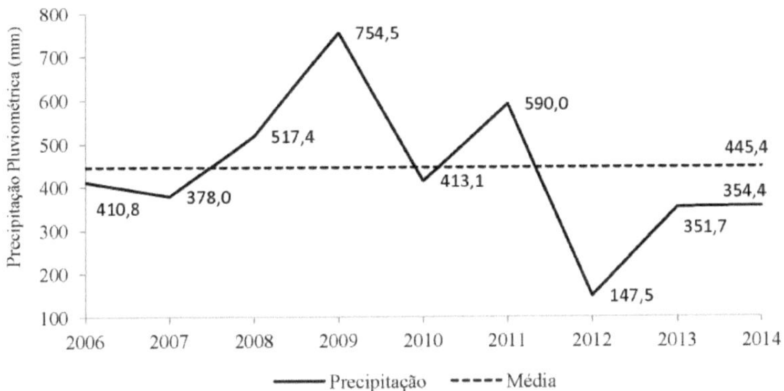

Graph 1 - Annual rainfall for the Petrolina - PE region (2006 to 2014)

Prepared by Elvis Pantaleão Ferreira (2015).

In the period between 2006 and 2014, 2012 was the most critical year in terms of rainfall (Graph 1), with a total rainfall of 26.86% of the historical average. Data published by MAPA (2013) emphasises that it was a year of severe drought throughout the semi-arid region, similar to remarkable episodes experienced in past decades. This situation disrupts the rural economy and significantly affects the entire population. However, in 2013, it rained a little more than twice as much as in 2012, still 36 per cent below the historical average. In 2014, when this research into water management in production cisterns was carried out, rainfall was slightly lower than the historical average.

higher than 2012 and similar to 2013, but still 64.55 per cent of the historical average.

With regard to the distribution of rainfall in 2014, the data shown in Graph 2 suggests that the study area shows intense spatial and temporal irregularity in rainfall. Graph 2 also shows that during the rainy season, April was the wettest month, when 73.5 mm of rain fell, followed by December during the dry season, when 79 mm were recorded, which for soil moisture purposes will contribute to agricultural activities in the following year (2015). It can also be seen that the number of days with rainfall is lower than the number of days without rainfall in all the months of the year, followed by an oscillating pattern, with a certain similarity between August and October.

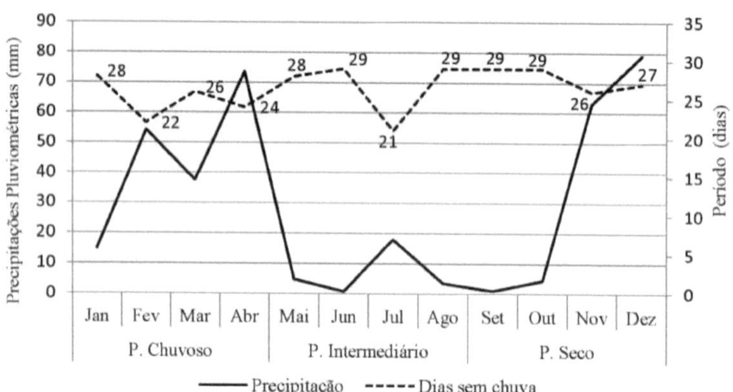

Graph 2 - Monthly distribution of rainfall and the number of days without rain for 2014 in the study area

Prepared by Elvis Pantaleão Ferreira (2015).

The rainy season between January and April (Graph 2) is responsible for the greatest amount of water in the soil, when the highest relative humidity levels are recorded, which vary on average from 66% to 73% (Teixeira, 2010). This contributes to a reduction in evapotranspirometric processes, favouring the maintenance of soil humidity and helping to reduce the frequency of crop irrigation, saving the water stored in the cistern for later use.

During the intermediate and dry periods, there is a need to apply a greater volume of water from the cistern to the crops, due to the low rainfall. Specifically, the dry season is characterised as the hottest period of the year, with temperatures ranging from 29.6 °C to 33.9 °C, which, combined with the low relative humidity of less than 55% (Teixeira, 2010), means that there is a greater frequency of irrigated days. There was an exception at the end of the dry season, when the highest rainfall in 2014 was recorded, in the order of 79 mm.

Regardless of the water used in an irrigated system, it is essential to assess its quality beforehand (Mantovani et al., 2013). Although the water used to irrigate the crops in this experiment came from rainfall, which is commonly considered to be of good quality, Souza et al. (2007) and Audry & Suassuna (2009) emphasise that it is important to know some indicators of its quality, given the possible implications, basically pH, which can occur due to pollution of the atmospheric compartment, from natural or anthropogenic sources, at times of rainfall, which is not expected in rural areas of the Brazilian semi-arid region.

Souza et al. (2007) comment that, for irrigation purposes, water monitoring is essential if the water source is located in originally saline soil, or in regions where evapotranspiration is very high, causing the concentration of salts present in surface water, or the source receives any industrial waste. Therefore, considering that rainwater is of superior quality (Jaques, 2005), we opted only to determine the pH, total concentration of soluble salts and the sodium adsorption ratio, in order to classify the water in terms of the danger of salinisation and alkalinisation of the soil.

The water used in this study had a pH of 6.75; a total concentration of soluble salts, also known as electrical conductivity (EC) of 0.06 dS m^{-1} at 25°C; and a sodium adsorption ratio (SAR) of 0.23 (Table 3). This means that the water has no restrictions for use by plants, according to the limits discussed and presented by Ayers & Westcot (1999). Therefore, in general, the water is classified as C1S1, considered to be of excellent quality for irrigation purposes (Bernardo et al., 2013). The C1 classification, regarding the danger of salinisation, indicates a low level of salts. The S1, regarding the danger of alkalinisation or sodification, indicates that the water has a low sodium concentration.

Table 3 - Analysis of the water stored in the cistern

	Parameters	**Acronym**	**Unit**	**Value**
Cations	Calcium	**2+ Ca**	mmol/L	0,41
	Magnesium	Mg^{2+}	mmol/L	0,29
	Sodium	In^{+}	mmol/L	0,21
	Potassium	K^{+}	mmol/L	0,09
	Sum		mmol/L	1,00
Anions	Carbonates	$CO3^{2-}$	mmol/L	0,00
	Bicarbonates	$HCO\ 3^{-}$	mmol/L	0,44
	Sulphates	SO4 **2-**	mmol/L	0,00
	Chlorides	Cl^{-}	mmol/L	0,45
	Sum		mmol/L	0,89
	pH	-	-	6,75
	C.E (25°C)	-	ds/cm	0,06
	Total Hardness	**CaCO3**	mg/L	3,48
	Sodium adsorption ratio	RAS	-	0,23
	Water classification	-	-	C1S1

Water was applied to the orchard increasingly, with the plants receiving different volumes in each period, always within the limits of 8, 12 and 16 litres per plant, applied three times a week in the rainy, intermediate and dry periods, respectively. Only the frequency fluctuated, as it was inversely proportional to the number of days with rainfall (**Error! Reference source not found.**).

The consumption of water applied to the orchard during the rainy, intermediate and dry periods was 5,760, 10,584 and 12,096 litres respectively (**Error! Reference source not found.**), totalling 28,440 litres of water applied to the fruit trees during 2014. The increasing volume of water applied was due to the increasing number of days without rain, culminating in a greater number of days that the plants received water stored in the cistern. The month with the lowest number of water applications was April, followed by December, when 73.5 mm and 79.0 mm of rain fell, respectively.

Table 4 - Rainfall (mm), number of days without rain and application of water to fruit trees.

PERIODS	RAINY				INTERMEDIATE				DRY				TOTAL
	Jan	Feb	Sea	Apr	May	June	Jul	Aug	Set	Out	Nov	Ten	-
Dias P.	3	6	5	6	3	1	10	7	1	2	4	4	47
Subtotal	20				16				11				
P (mm)	14,9	54,2	37,6	73,5	4,7	0,7	17,9	3,5	1	4,4	63	79	354,4
Subtotal	180,2				26,8				147,4				
No. of days applied Water	12	10	11	7	12	13	11	13	13	14	9	6	131
Subtotal	40				49				42				
L period	5.760				10.584				12.096				28.440

Days P = days with precipitation: P (mm) = precipitation; N° days applic. Water = number of days with water application; L period = litres of water applied per period.

It should be noted that the 354.4 mm of rainfall that occurred in 2014 during the experiment (Graph 1) made it possible to contribute approximately 708.8 litres of water per plant, considering that each plant has a micro-basin of approximately 0.8 m in radius. This allowed rainwater to be retained in each plant. Adding the rainwater to the 1,896 litres of water from the cistern, each plant received an average of 2,604.8 litres per plant during 2014.

The volume of water stored in the cistern and used in the beds is shown in (**Error! Reference source not found.**). For bed 1, 3,424, 3,680 and 3,456 litres were used during the rainy, intermediate and dry periods respectively, and in the second bed, 50% of these values were used, totalling 15,840 litres of water, around 10% less than expected (Table 2), given the occurrence of days without irrigation, spared by the incidence of significant rainfall. The maximum volume of water adopted for each water application per bed was 32 and 16 litres per day. This volume corresponded to 8.0 mm and 4.0 mm for beds 1 and 2 respectively, given that both have dimensions of 4 m .[2]

The total volume of water applied to beds 1 and 2 in the dry period was slightly lower (6 per cent) than in the intermediate period. This was due to the heavy rainfall at the end of the dry period, during November and December. On this occasion there was a reduction in the frequency of irrigated days, which meant that at the end of the dry period the total volume of water applied to the beds was lower than the volume applied in the intermediate period.

44,280 litres of water were applied to the crops, 64.22% in the orchard and 35.77% in the beds. There was therefore a balance of 7,720 litres of water inside the cistern, which corresponds to 273.1 mm of water in the cistern. This volume is important for maintaining a reserve (dead volume) so that in the event of a possible scenario of a prolonged dry period, there will be water to keep the crops alive, as well as compensating for any losses due to evaporation and infiltration and losses due to the use of water by the farmer.

Diaconia (2008), in his book "Construction of a 52,000-litre pavement cistern", comments that it is important to always keep water in the cistern in order to maintain the structure's humidity and avoid physical damage such as the appearance of micro-cracks in the joints of the plates, due to the expansion and contraction processes caused by the temperature difference, which can lead to a loss of water containment capacity.

Table 5 - Rainfall (mm), number of days without rain and water applied to beds 1 and 2

PERIODS	RAINY				INTERMEDIATE				DRY				TOTAL
	Jan	Feb	Sea	Apr	May	June	Jul	Aug	Set	Out	Nov	Ten	-
Days P	3	6	5	6	3	1	10	2	1	2	4	4	
Subtotal	20				16				11				47
P(nnn)	14,9	54,2	37,6	73,5	4,7	0,7	17,9	3,5	1	4,4	63	79	354,4

Subtotal	180,2				26,8				147,4				
N® days aphc. water	28	25	30	19	27	30	27	31	30	31	25	21	324
Subtotal	102				115				107				
Lcant. 1	896	832	960	736	864	960	864	992	960	992	800	704	10.560
Subtotal	3.424				3.680				3.456				
L cant. 2	448	416	480	368	432	480	432	496	480	496	400	352	5280
Subtotal	1.712				1.840				1.728				
Total (L)	5.136				5.520				5.184				15.840

DaysP = days with precipitation; P (mm) = precipitation; N° days aphc. Water = number of days with water application: L period = litres of water applied per period. L cant. 1 and 2 = litres of water applied to beds 1 and 2.

4. 2Chemical and physical characteristics of the soil in the experimental area

4.2. 1Soil chemical attributes

Soil fertility is one of the most important components of production systems, especially for crops such as fruit and vegetables. According to Prezotti & Guarçoni (2013), it is one of the essential parameters for sustainable agriculture. Table 4 shows the chemical characterisation of the soil in the experimental area. The high, medium and low reference classes used for interpretation were based on the soil analysis interpretation tables for the state of Pernambuco (Cavalcanti, 1998).

Table 4 - Chemical characterisation of soil samples

Professor	pH	Al^{3+}	Ca^{2+}	Mg^{2+}	H+Al	K	P	In	M.O
m	**(CaCl2)**	-----	cmolc dm $^{-3}$-----			---	mg dm $^{-3}$ -		dag Kg^{-1}
0,0 - 0,2	6,1	0,0	2,5	0,6	1,6	150	20	61	1,4
0,2 - 0,4	6,0	0,0	1,9	0,4	1,6	140	7	46	1,0
Professor	SB	t		T	m		V		PTS

m	---------- cmolc dm $^{-3}$ --------			----------------- % -----------------		
0,0 - 0,2	3,5	3,5	5,1	0	68,5	5,2
0,2 - 0,4	2,7	2,7	4,3	0	62,4	4,7

SB: sum of exchangeable bases; CTC (t) = effective cation exchange capacity; CTC (T) = cation exchange capacity; m = aluminium saturation index; V = base saturation index; PTS = percentage of exchangeable sodium, also known as the soil sodium saturation index.

The values for active acidity (pH) and exchangeable acidity (Al^{3+}) at the depths of 0.0 to 0.2 and 0.2 to 0.4 metres are classified as weak acidity and low exchangeable acidity, respectively (Table 4). It should be noted that the pH was determined in CaC12, which is considered by Davey & Conyers (1988) apud. Ebeling et al. (2008) to be a more accurate determination than pH in water, since the activity of the ion in the soil solution is less affected by salts present in the soil.

Potential acidity (H+Al) is low and does not limit crop development, probably due to the low active acidity and the absence of exchangeable aluminium Al^{3+} in the soil solution. These conditions are favourable for plant development as they do not cause crop toxicity (Horst et al., 2010), given that aluminium cations in the soil solution, especially in inorganic monomeric forms, negatively affect nutrient absorption and root growth (Echart & Molina, 2001).

Calcium levels were classified as medium for both depths, with levels ranging from 2.5 to 1.9 cmolc dm^{-3} , the latter being attributed to the 0.2 to 0.4 m layer (Table 4). The lower calcium value found in the deeper layer can be attributed to its consumption by the crops. Meurer (2006) argues that a deficiency of this element is usually found in acidic soils, a characteristic not observed in the soil surveyed. Klaus (2007) comments that as well as playing an important role in the growth of roots and shoots and increasing tolerance to heat, wind and cold stress, calcium also improves the structure, permeability and infiltration of water in the soil.

The magnesium levels obtained were slightly low, ranging from 0.4 to 0.6 cmolc dm^{-3} , with a Ca:Mg ratio of 4:1 for both depths (Table 4). Data presented by Prezotti et al. (2007) and Salvador et al. (2011) show that most crops have a good nutritional balance when the Ca:Mg ratio varies between 3:1 and 4:1. Among other functions, Malavolta (2006) and Vitti et al. (2006) emphasise that magnesium is an essential enzyme activator for energy transfer in the photosynthesis process.

Phosphorus levels decreased with depth, with the highest value of 20 mg dm-3 in the first layer of 0.0 - 0.2 m, classified as low (Table 4). This characteristic is inherent to the soil studied. This behaviour can be attributed to the low mobility of this element, which is concentrated in the topsoil, or it can be associated with the formation of low-solubility calcium phosphate precipitates (Rheinheimer & Anghinoni, 2001). Corrêa et al. (2011) point out that low phosphorus levels characterise most soils in tropical regions, where the clays have a high capacity for phosphorus adsorption, which makes the soil a competitor for the plant, together with its low solubility.

Potassium levels were average for both depths (Table 4). Like nitrogen, plants require a large amount of potassium, which is important in several essential processes, such as enzyme activation, photosynthesis, water regime, amino acid formation and protein synthesis (Malavolta, 2006).

The sodium value, only in the most superficial layer, can be classified as slightly high as it showed a concentration of more than 50 mg dm^{-3} (Table 4). This may be associated with the high evaporation rates and low rainfall, promoting the capillary rise of salts to the soil surface as the water is evaporated or consumed by the plant, as well as the mineralogical characteristics of the soil (Mantovani et al., 2013). According to Pereira (1998) and Mota & Oliveira (1999), the soils of the Brazilian semi-arid region, due to the characteristics of their climate, relief, geology and drainage, often present favourable conditions for the occurrence of a predominance of sodium and other salts. According to Erthal et al. (2010), high concentrations of Na^+ in the soil solution can cause deterioration of the soil structure by dispersing the colloids and subsequently clogging the macropores, causing a decrease in permeability to water and gases.

With regard to organic matter (Table 4), the levels ranged from 1.4 to 1.0 dag Kg^{-1} depending on depth, classified as low. However, a higher content was observed in the surface layer, close to the upper limit of this class of 1.5 dag Kg^{-1} . Although there is a continuous supply of mulch in this area, due to the accumulation of coconut bagasse in the projection of the plant canopy, it is believed that the low MO contents are related to the environmental conditions of the region, particularly temperature and humidity, which directly interfere with the decomposition rate of this material (Cardoso et al., 2013).

The soil's sum of exchangeable bases (SB) and effective cation exchange capacity (t) values were classified as medium for both depths, ranging from 2.7 to 3.5 cmolc dm^{-3} (Table 4). The cation exchange capacity (T) showed a decrease as a function of depth, classified as medium to low, respectively, at the depths studied. T is a characteristic of the soil and can only be altered by the addition of organic matter or as a result of erosion processes, in which case the surface layer is lost (Prezotti et al., 2007). Therefore, as the soil in this area has a higher (T) in the surface layer, it is possible that the addition of coconut bagasse in the projection of the plant canopy has contributed to an increase in the organic matter content, which reflects a better condition for retaining cations, combined with the fact that the site does not have active and exchangeable acidity in high concentrations.

Base saturation (V) greater than 50% (Table 4) indicates that the soil is eutrophic in nature and is characterised as having good natural fertility due to its high base value, especially calcium Ca^{2+} , which has a positive influence on reducing soil acidity. The soil has no problems in terms of aluminium saturation or sodium saturation. According to Amorim et al. (2010), the soil in the study area was classified as non-sodic for both layers in terms of sodicity as a function of PST.

According to the data presented, this is an area that does not offer limitations to crop development, although it does have some attributes that are slightly below the reference values. Considering that water is a limiting factor for crop development, the aim of this study is not to achieve maximum crop yield potential given the characteristics and limitations of the soil and water, but rather to reflect the conditions experienced by family farmers in the Brazilian semi-arid region.

4.2. 2Physical attributes of the soil

4.2.2.1 Physical characterisation of the soil

The results of agricultural production are essentially dependent on the characteristics of the climate, the soil and the plants, as well as the relationships between them. This interaction is due to the fact that the soil is the main substrate for plant support and fixation, as well as serving as a water reservoir and source of nutrients that are essential for crop development. Table 5 shows the specific granulometric classification.

Table 5 - Soil particle size classification

Professor	Granulometry						Textural Class
	Sand*				Silt	Clay	
	G	M	F	Total			
m	----------------------			%	----------------------		--
0,0 - 0,2	45,3	24, 3	10,8	80,4	7,8	11,8	Sandy loam
0,2 - 0,4	24,0	16, 4	27,0	67,4	8,1	24,5	Average

1G: coarse; M: medium; F: fine.

According to the granulometric composition of the soil at the depths investigated 0.0 - 0.2 and 0.2 - 0.4 m, the soil texture was classified as sandy loam and medium, respectively (Table 5). The reduction in sand fractions and increase in fine fractions observed in the 0.2 to 0.4 m layer, especially in the clay fraction, showed an increase of around 50% in relation to the top layer, providing a greater number of pores and, consequently, a greater capacity for retaining and storing water in the soil profile.

This is favourable for the soil cultivation system, since water is one of the most limiting factors. In addition to this characteristic, the presence of a higher clay fraction may favour greater nutrient adsorption over time, as well as crop development in the drier months of the year, commonly between September and December.

The amount of sand fraction greater than 80% recorded in the 0.0 to 0.2 m layer (Table 5) shows that it is permeable and therefore retains little water, allowing water to pass quickly through the pores and then reach the deeper layers, favouring the loss of nutrients through leaching. Therefore, sandy-textured soils commonly have water limitations due to their greater drainability (Klein, 2008; Macedo, 2014).

In view of this textural characteristic inherent to the soil, granulometric analysis has been used by the Ministry of Agriculture, Livestock and Supply - MAPA (2008) as a technological tool that guides the granting of funding to rural producers, in view of agricultural zoning for climate risk. It also helps with proper soil conservation management, irrigation management and fertiliser management, among other things.

As for the degree of flocculation of the clays shown in Table 6, the 0.2 to 0.4 m soil layer shows a better degree of aggregate stability than the top layer, with a degree of flocculation of more than 20% and a natural clay content of less than 9%. The lack of mechanical tillage, combined with the presence of organic matter (Table 4), may have contributed to the low levels of natural clay, promoting greater aggregation of the particles due to the cementing action of the organic matter (Prado & Centurion, 2001). Studies carried out by Donagemma et al. (2003) describe that low natural clay content contributes positively to the non-sealing of part of the soil's subsurface pores, preventing an increase in density, a decrease in permeability to water, gases and the infiltration rate.

Table 6 - Physical characterisation of the soil

Professor	**Clay Natural**	**Degree of Flocculation**	**CEes 1:2,5**	**Density**		**Total porosity**
				Soil	Particles	
m	g Kg^{-1}	%	µS cm^{-1}	--- Kg dm ^{3}		%
0,0 - 0,2	8,81	24,50	50,90	1,47	2,57	42,80
0,2 - 0,4	8,72	20,20	47,80	1,42	2,56	44,53

Natural Clay = ADA (water dispersed clay); **ECes** = electrical conductivity of the saturation extract.

The surface layer from 0.0 to 0.2 m showed a slight increase in natural clay (Table 6). This behaviour may be associated, above all, with the accumulative effect of sodium and potassium in this layer (Table 4), since these elements are considered important dispersants (Homem et al., 2012) and can cause the dispersion of clay in the soil and contribute to the disintegration of particles with the impact of water droplets on the surface (Donagemma et al., 2003).

Although the region has an evaporation rate that exceeds the annual rainfall (Teixeira, 2010), it promotes the capillary rise of salts to the soil surface as the water is evaporated or consumed by the plant. Analysis of the concentration of soluble salts (ECes) showed low values of soluble salts in both layers (Table 6), characterising them as non-saline soil, as discussed by Oliveira et al. (2002). The low concentration of soluble salts is important because it doesn't interfere with the plant's absorption of water and nutrients. If this concentration were high, it would cause an increase in the osmotic pressure of the soil solution, which could lead to physiological drought and nutritional imbalance (Oliveira et al., 2002), as well as sealing the soil surface and forming impermeable layers in the subsurface (Barros et al., 2006).

As far as soil density is concerned, there is a decrease as a function of depth. This behaviour is associated with the predominance of fine-textured particles in the soil, such as silt and clay (Table 5), resulting in lower soil density and an increase in total porosity and, consequently, greater water retention and storage capacity.

Generally speaking, it is important to emphasise that family farmers who have a production cistern and who cultivate vegetable gardens and orchards in soils with better physical and chemical conditions than those presented in this study (Table 4 and Table 5) will be able to provide greater water storage capacity in the soil, positively influencing soil humidity and consequently having better growing and production conditions.

4.2.2.2 Soil Water Retention Characteristic Curve

Given the sandy/medium texture of the soil (Table 5), a tension of 10 kPa was adopted for Field Capacity (FC). According to White (2005) and Bernardo et al. (2013), FC is characterised by the maximum water content in the soil retained against gravitational force. Graph 3 shows the Soil Water Retention Characteristic Curves (SWRC) for the 0.0 to 0.2 and 0.2 to 0.4 metre layers.

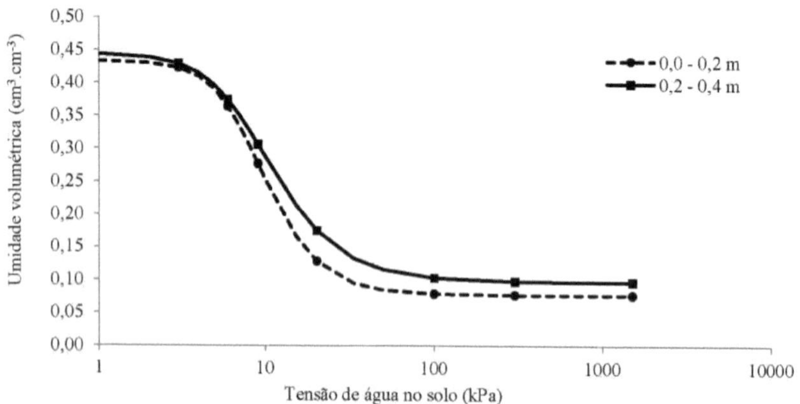

Graph 3 - Soil Water Retention Characteristic Curves (SWRC) for the 0.0 - 0.2 and 0.2 - 0.4 metre layers

It was found that the saturation humidity for the two depths of 0.0 - 0.2 and 0.2 - 0.4 metres showed values close to 0.45 cm^3 cm^{-3} , with water contents at Field Capacity (FC) of 0.2565 and 0.2859 cm^3 cm^3 and at Permanent Wilting Point (PWS) of 0.0772 and 0.0973 cm^3 cm^{-3} , respectively for these layers.

The Soil Water Retention Characteristics Curves (SWRCs) for both depths tend to be similar (Graph 3). However, there is a slight decrease in the soil's volumetric humidity, which corresponds to the entry of air into the macropores, related to the predominance of the soil fraction composed of sand, corresponding to around 80% for the 0.0 to 0.2 m layer and 70% for the 0.2 to 0.4 m layer (Table 5). At approximately 3 kPa, the inflection point begins, at which point the maximum suction force is sufficient to begin draining water from the macropores (Fredlund & Xing, 1994; Gerscovic, 2001).

In Graph 3, from 3 kPa up to around 100 kPa there was a sudden loss of water content in the soil as the tension increased. According to Miguel et al. (2006), this is a typical characteristic of coarse-textured soils (Table 5), where the water retained in the macro and meso pores drains rapidly as the tension increases.

In the CCRAS for the 0.2 - 0.4 metre depth, it was found that, from tensions greater than 10 kPa, there is more water retained for the same tension values than the top layer, with a contribution of just over 34.59% more available water than the curve for the 0.0 - 0.2 m depth. This characteristic is associated with the greater quantity of fine fractions in this soil layer, mainly clay (Table 5).

An increase in these fractions generally leads to an increase in the amount of water retained under the same stress value, due to the greater number of pores and, consequently, greater water storage capacity in the soil. Studies carried out by Miller et al. (2002) and Reichardt & Timm (2012) presented data confirming this behaviour.

Crops with root systems that exploit the top 0.0 to 0.2 metres of soil tend to have difficulty absorbing water and adsorbing nutrients. In addition, this layer is subject to greater moisture loss through evaporation as it is more exposed, unlike the underlying layers. For this reason, when the topsoil is cultivated with vegetables, there is a need for more frequent application of replacement water than crops with root systems that exploit the soil in depth.

At the lower limit of the curve, it was observed that from tensions greater than 1000 kPa there is an insignificant behaviour in the moisture values at both depths as the tension increases, this point in the curve being defined as the residual moisture content (Gerscovich, 2001). Fredlund & Xing (1994) define residual moisture content as the soil moisture content from which a large variation in tension is required to remove insignificant variations in moisture content.

Soils with a predominance of the clay fraction, because they retain more water, can offer better conditions for cultivation systems that use cisterns, since water is one of the most limiting factors. On the other hand, sandy/medium-textured soils have a low water retention capacity and a high evaporative demand (Urach, 2007). Considering that the area of the experiment had a sandy texture, the strategy adopted to control soil humidity and temperature was to keep the plant canopy protected by mulching with coconut bagasse, in order to minimise water loss and keep the soil moist for longer.

Data published by Bertoni & Lombardi Neto (2008) highlights that mulching not only helps to maintain soil temperature and humidity, but also reduces soil disintegration, reduces the incidence of weeds and improves the supply of organic matter, which retains around three times its weight in water.

4.3 Water content in the soil profile

4.3.1 Orchard area

The results of the volumetric moisture in the soil profile of the experimental units of the treatments that received water from the cistern (Table 1), for the rainy, intermediate and dry periods, respectively, and of the treatments that only received rainwater, will be presented and discussed below.

For the orchard area, data from the Characteristic Soil Water Retention Curve (CCRAS) was adopted for the 0.40 m layer, showing moisture at Field Capacity (CC) of 0.2859 $cm^3\ cm^{-3}$ and at Permanent Wilting Point (PMP) of 0.0973 $cm^3\ cm^{-3}$, considering that a large part of the root system of fruit plants is at this depth (Pires et al., 1999), We also discussed the lower limit of water availability in the soil - factor (f) equal to 0.7 (Bernardo et al., 2013), which corresponds to around 50% of Field Capacity, i.e. the minimum humidity that fruit trees can be subjected to without significantly affecting their production.

A) Rainy season

A1. Treatment with irrigation

Graph 4 shows that during the Rainy Period - CP, between January and April, the average soil moisture at 0.10 m depth remained at around 80% of Field Capacity - FC, ranging from 0.2064 to 0.2584 $cm^3\ cm^{-3}$. At the same time, with increasing depth there was an increase in soil humidity, up to around 0.30 m, when the soil reached its maximum humidity at around 90% of Field Capacity. In this case, it is possible that due to the sandy texture of the soil (Table 5), there is a guarantee of a greater flow of water and, consequently, a greater infiltration gradient, where the force of gravity prevails against the action of capillarity, favouring the flow of water to the deeper layers, which can also move according to gradients of other potentials that may be present (Libardi, 2005).

This process is called water redistribution (Loyola & Prevedello, 2006) and is characterised by an increase in moisture in deeper layers at the expense of water contained in the initially moistened surface layers. Graph 4 also shows a defined soil moisture profile, formed by the sinuous line at a depth of 0.10 to 0.40 metres. This line can be attributed to the movement of water in the soil, forming the so-called wet bulb, commonly observed in crops subjected to localised water application, as occurred in this experiment for the treatments that received water from the cistern.

It should be noted that the area wetted by the bulb corresponds to a larger area than that visualised on the soil surface, in this case reflecting the lower humidity on the surface, corroborating the data presented by Pires et al. (1999), who pointed out that the wetted area corresponds to the 0.10 to 0.20 m range, which can extend to a depth of 0.30 m, where the largest wetted diameter is normally located. It should be emphasised that soils with a non-uniform profile, as is the case with the soil in the current study, generally show greater lateral water movement than the same soil with a uniform profile (Bernardo et al., 2013).

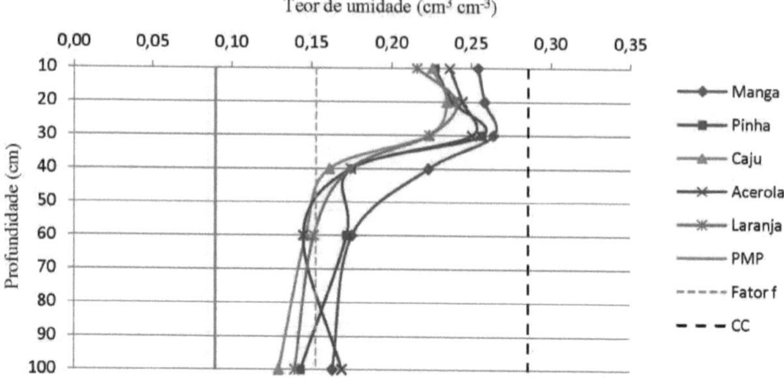

Graph 4 - Soil moisture profile with irrigated fruit trees during the rainy season

The redistribution of water in the soil occurs in all directions, with greater or lesser ease, with the shape and size of the wet bulb depending on the characteristics of the soil, the volume of water applied and the frequency (Libardi, 2005). However, in coarse-textured soils, the vertical movement of water predominates and the bulb is more elongated when compared to the same amount of water applied to clay-textured soils (Cook, 2006; Maia & Levien, 2010).

The high soil humidity observed during the rainy season, when compared to the intermediate and dry periods, is associated with the water coming from the rainfall that occurred, which corresponded to 180.2 mm, increased by the supply of 8.0 litres per plant applied by dripping three times a week, which was equivalent to a volume of water supplied of 5,760 litres, applied during this period.

It is also possible that the milder climate, with lower average temperatures, higher relative humidity and calmer winds, led to a lower evaporation rate during the rainy season, thus allowing for a higher water content in the soil. Climatological data, published by Teixeira (2010), shows that during the months of January to April, the Petrolina - PE region has the highest relative humidity, ranging on average from 66% to 73%; the lowest average wind speed values, ranging from 1.6 to 1.8 m s^{-1} and an average temperature between 26.9 °C and 26.2 °C.

It is also believed that the higher soil moisture values recorded specifically for the mango crop (Graph 4), which received the same amount of water as the other species, may be associated with the architecture of the plants, which, because they have a larger canopy and leaf area than the other crops, provide more shaded area, minimising evaporative processes. The opposite situation was observed for the orange tree.

The increasing variation in moisture retention in the soil profile, recorded up to 0.30 m, is possibly associated with the greater concentration of the soil's clay fraction in the 0.20 to 0.40 m depth, at around 50 per cent in relation to the top layer (Table 5). As well as the presence of the branched root zone, characterised by a more woody consistency, where there is a lower rate of water and nutrient absorption (Cintra et al., 1996; Konrad et al., 2001). Also according to the authors, when studying the development and distribution of the root system in fruit plants, they observed that the branched region of the roots is largely responsible for fixing the plants to the soil.

The decrease in soil humidity observed between 0.30 and 0.40 metres deep is probably associated with the root absorption zone, which means that the root system of fruit plants is exploiting more of the soil profile, contributing to the increasing reduction in soil water content (Graph 4). Work carried out on fruit plants by Cintra et al. (1996) and Soares et al. (1998) shows that the hairy zone of the roots, located above the distension zone, where the roots show the highest growth rate, is responsible for the highest percentage of water and nutrient absorption, given that the water potential of the root hairs is lower than that of the soil water, causing water to enter their interior.

From 0.60 m to 1.0 m depth, there was very little variation in humidity, with practically constant performance, with the exception of the pine tree and mango crops, which showed a decline in soil humidity. However, the moisture content in the mango tree remained higher throughout the soil profile. This behaviour is possibly associated with the very small number of roots in search of water. With regard to the critical humidity represented by the line corresponding to the f factor, characterised as the minimum humidity that crops can be subjected to without compromising production, this remained higher up to 0.40 m depth for all crops, with the greatest depth persisting for mango and pine crops.

A2. Treatment without irrigation

Graph 5 shows that the average soil humidity for the treatment in which the fruit trees did not receive water from the cistern, but only from rainfall, is lower throughout the soil profile when compared to the treatment in which the fruit trees also received water from the cistern during the same period. It should be noted that the average humidity recorded in the soil profile refers only to the 180.2 mm of rain that fell during this period.

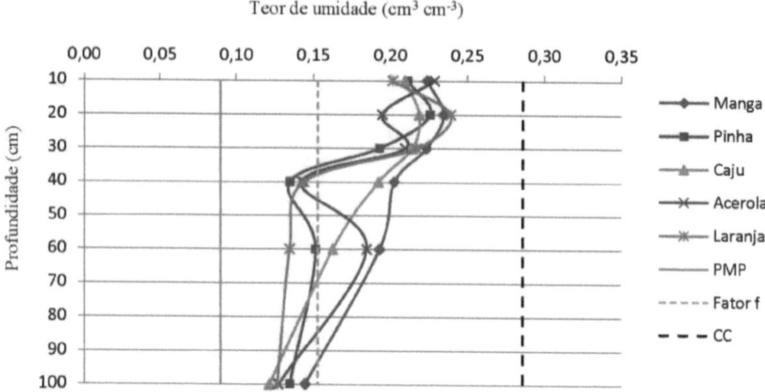

Graph 5 - Soil moisture profile with non-irrigated fruit trees during the rainy season

Similarly to the previous situation (Graph 4), with the exception of the surface layer, the moisture content in the mango tree also remained higher, as shown in Graph 5. A particular characteristic observed in this treatment, whose fruit trees did not receive water from the cistern, is for the acerola crop, where there was a continuous loss of moisture in the soil up to 0.20 metres. This may be associated with the greater consumption of water by the plant in the 0.10 to 0.20 m interval, as well as variations in the amount of sand in the soil, promoting rapid internal drainage of water and reflecting a greater contribution of moisture in the next layer.

Between the 0.30 and 0.40 m intervals, there is a marked loss of moisture, most noticeably for the pine, acerola and orange crops, which may be associated with the higher root density zone and consequent greater demand for water. From 0.60 m there is a marked decrease in soil humidity, characterised by the small amount of water that reaches this depth, converging at 0.1349 cm^3 cm^{-3} at 1.0 m depth for most crops. Figure 9 shows a partial view of the study area during the rainy season.

Figure 9 - Partial view of the orchard during the rainy season

Photo: Nilton de Brito Cavalcanti (2014).

B) Intermediate Period

81. Treatment with irrigation

During this period, the average soil humidity for the mango and pine tree crops was 15% above the critical humidity between 0.20 and 0.50 m deep (Graph 6), while the other crops were below. During this period, 12 litres of water per plant were applied by dripping three times a week, which corresponded to a total of 10,584 litres. Even so, this was not enough to prevent the soil humidity for most of the crops from remaining around 20% above the Permanent Wilting Point at a depth of 0.10 metres.

The marked loss of moisture observed in the most superficial part of the soil profile, notably between 0.10 and 0.20 m, is attributed to the lower rainfall rate between May and August, which comprises the intermediate period. During this period, only 26.8 mm of rain was recorded, which corresponds to only 15% of the rainfall during the rainy season, associated with a decrease in relative humidity ranging from 65 - 60% (Teixeira, 2010) and the sandy texture of the soil, which favours an increase in the rate of water vapour transfer to the atmosphere, due to the air heated on the soil surface during the hottest hours of the day.

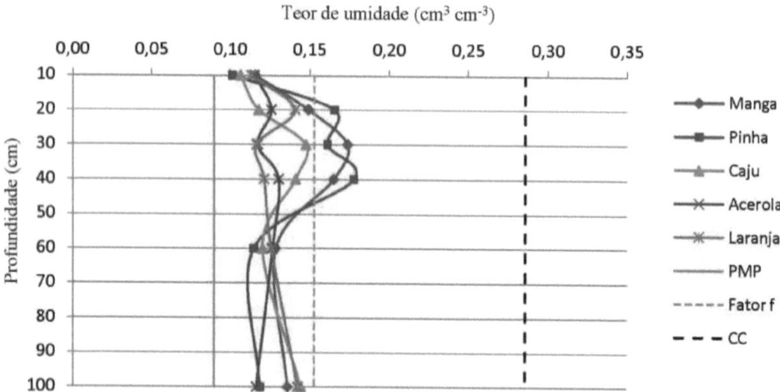

Graph 6 - Soil moisture profile with irrigated fruit trees during the interim period

The graph above shows that from approximately 0.15 and 0.20 metres deep, the humidity in the soil profile for the pine tree and mango crops is above the critical humidity, persisting until 0.47 metres deep, indicating the possibility that even at this time, which corresponds to the intermediate period between May and August, these crops were facing soil humidity values that, if they remained, could produce without significantly affecting production. It is possible that spatial changes in the soil's physical and water properties, as well as the size of the crop's leaf area and its relationship with the size of the shaded area, may have led to this behaviour.

At a depth of 0.30 metres, the average soil humidity for orange and acerola crops decreased considerably compared to the other crops, following an increase in soil water content in the previous layer. This behaviour may indicate that the root system of these crops, faced with low water availability, also explores the more superficial layers of the soil in search of water. Research carried out by Coelho et al. (2002) and Santos et al. (2005) with fruit trees subjected to water deficits shows that the root system develops towards the surface layers, between 0.10 and 0.30 m, influenced by the redistribution of water in the soil profile.

A slight increase of around 15 per cent in average soil moisture can be seen in Graph 6, at a depth of 1.0 m, compared to the rainy season, which may be associated with the contribution of the moisture content from the previous period plus the intermediate period, thus increasing soil moisture over time.

In this sense, studies published by Reichardt & Timm (2012) show that even when rain or irrigation ceases, the movement of water within the soil profile can persist for days or months, ceasing due to physical impediments in the soil or when matric potential becomes more important than gravitational potential as the soil loses water. However, great importance must be given to the most superficial layers, where there is the maximum concentration of roots and the greatest demand for water to meet the plants' needs.

It should be noted that under the conditions studied, the volume of water applied from the cistern, plus rainfall, does not meet the plants' needs. However, the aim is to maintain the fruit trees with a minimum water requirement and to obtain production, even on a small scale, in order to promote the inclusion of fruit and vegetables in the diet of rural families in the Brazilian semi-arid region.

82. Treatment without irrigation

The behaviour of the average soil moisture of the treatment in which the plants did not receive water from the cistern is shown in Graph 7. At a depth of 0.10 m, it shows average soil moisture values practically common to all crops, in the order of 0.09667 $cm^3\ cm^{-3}$, corresponding to 7% above the Permanent Wilting Point. This behaviour is also observed up to the 0.20 m layer, although there is a slight increase in moisture. From the 0.20 m depth onwards, average soil moisture values slightly above 20% of the Permanent Wilting Point are recorded for all crops, with the exception of mango and acerola.

From 0.40 m to 1.0 m deep (Graph 7), the soil moisture lines show little oscillation, reflecting less sinuosity in the moisture curves when compared to the treatments that received water from the cistern. It can also be seen that all the moisture lines for water extraction by the crop roots remain below the minimum moisture, a characteristic associated with not applying water to this treatment.

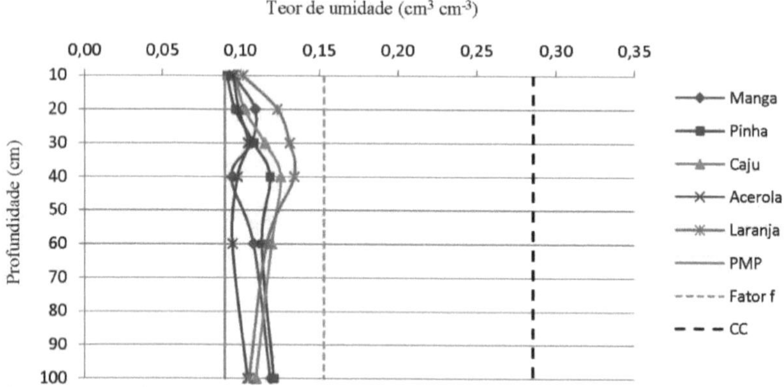

Graph 7 - Soil moisture profile with non-irrigated fruit trees, intermediate period

C) Dry Period

C1. Treatment with irrigation

The humidity levels for the treatment whose plants received water from the cistern during the dry period, which includes the months of September to December, can be seen in (Graph 8). During this period, a total of 147.4 mm of rainfall was recorded, with the greatest concentration of rainfall, around 95%, in the months of November and December, 63 mm and 79 mm respectively, concentrated in just 4 days, as shown in Graph 2.

On this occasion, there were the highest number of days when water was applied to the fruit trees, with the first two months of this four-month period standing out. In the last period of the studies, the crops received 16 litres of water per plant, three times a week, totalling 12,096 litres at the end of the four-month period, applied in a similar way to the previous periods. Of this total volume, 75 per cent was applied in the first two months and the rest in November and December.

Graph 8 shows that even though 16 litres of water were administered per plant at a distance of 0.40 m from the stem, the results obtained indicate that the behaviour of the average soil moisture in the top layer showed low soil moisture values for all crops, around 10% above the Permanent Wilting Point - PMP. However, there was a gradual increase in soil moisture up to a depth of 0.20 m, varying from 0.1087 cm^3 cm^{-3} to 0.1543 cm^3 cm^{-3} , which is possibly associated with the following factors: the amount of water applied to the crops; the volume of rainfall recorded at the end of this period; and the greater concentration of the soil's clay fraction at a depth of 0.20 to 0.40 m, around 50% in relation to the top layer (Table 5).

Although there was a significant increase in soil moisture in the subsurface layer, it was not enough to reach above the minimum moisture level (0.153 cm^3 cm^{-3}). According to Coelho et al. (2008), under these conditions plant growth can be affected, causing a reduction in production. A distinct increase in soil humidity was also observed for the mango crop, followed by the cashew crop, a characteristic also recorded in the previous periods. In addition to the architecture of the plants, this could possibly be due to the unevenness of the area or to spatial variations in the soil's physical and water properties, causing the precipitated water to flow towards these crops.

In the following layers up to 0.60 m, a decline in soil moisture is observed, followed by an intersection of the moistures at 0.60 m, at 0.1158 cm^3 cm-3, showing around 20% above the Permanent Wilting Point - PMP, with practically uniform behaviour up to 1.0 m.

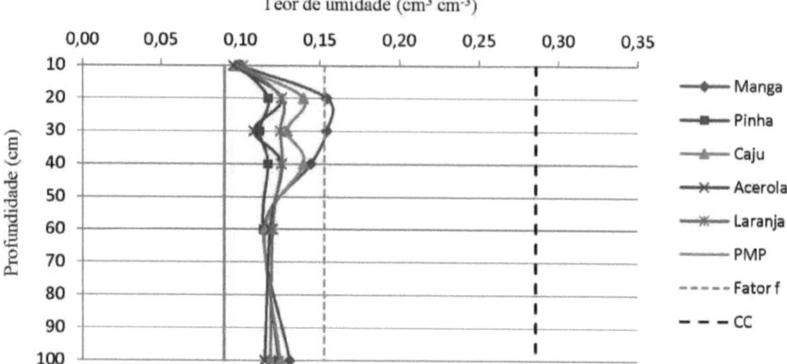

Graph 8 - Soil moisture profile with irrigated fruit trees during the dry season

The low moisture content, especially in the top layer of the soil and in depth, seen in Graph 8, may be associated with climatological factors that affect this period of the year, according to agrometeorological information published by (Teixeira, 2010), and therefore compete for the amount of water added to the soil. Among the factors climatological factors affecting this period include the lowest relative humidity recorded for this period, below 55%, and the highest wind speeds, twice as high as during the rainy season, reaching 3.3 m.s $.^{-1}$

This period of the year has the highest temperatures, ranging from 26.5 °C to 28.1 °C, and potential evapotranspiration reaching 166.67 mm. These conditions, combined with the sandy/medium texture of the soil, with around 80% of the sand fraction recorded in the 0.0 to 0.20 m depth, have contributed to the intensification of the flow of water vapour from the soil to the atmosphere.

As for the decrease in moisture values observed from 0.40 m onwards (Graph 8), it is possible that this is associated with the greater density of the root system around this layer and its ability to intercept and absorb moisture from the soil, as well as the dissolved elements that are carried by the water by mass flow, thus reducing the advance of moisture into the underlying layers. It is also possible that the volume of water applied to the plants, together with the contribution of rainfall during this period, was insufficient to reach the deeper layers of the soil. Nascimento et al. (2012)

described soil moisture behaviour similar to that found in this study, evaluating soil moisture in fruit plants at 0.40 m from the stem, irrigated with water stored in a P1+2 cistern, under similar conditions to this research.

C2. Treatment without irrigation

The behaviour of the average soil moisture for the treatments that did not receive water from the cistern during the dry period is shown in Graph 9. For all crops, there were low soil moisture values, very close to the Permanent Wilting Point - PMP, persisting up to 0.20 m, when compared to the same plants that also received no water in the previous period.

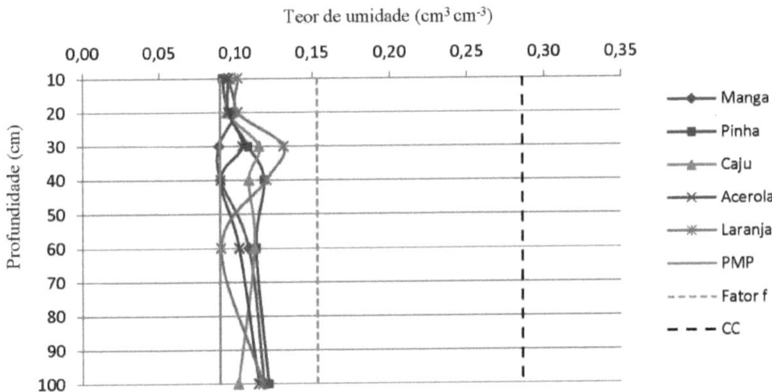

Graph 9 - Soil moisture profile with non-irrigated fruit trees during the dry season

Despite the low soil moisture levels, the graph shows higher moisture values between 0.20 and 0.40 m depth, which may be associated with the presence of a greater amount of the soil's clay fraction (Table 5), ranging from 11.82% in the 0.0 to 0.20 m layer to 24.51% in the 0.20 to 0.40 m layer, providing greater water storage capacity in the soil. The very low soil moisture values observed for these plants are associated with the climatic factors of that period and the lack of water applied to the crops in this treatment.

Unlike the treatment that received water from the cistern, during this period the moisture content of the soil explored for the mango tree showed moisture values equal to the Permanent Wilting Point. This species behaved differently to the other crops, which may be related to the high demand for water, and possibly this condition meant that the mango plant was unable to resist (Figure 10B). With regard to soil moisture at 1.0 m, there was practically no change in behaviour compared to the plants in the intermediate period, which also did not receive water from the cistern.

Figure 10 - Partial view of the orchard during the dry season: (A) mango plant that received water from the cistern; (B) mango plant that did not receive water from the cistern.

Photo: Nilton de Brito Cavalcanti (2014).

One factor that may be reflecting the low moisture levels observed during the experiment may be the measurement of the water content in the soil obtained by the sensor, as the range is only 0.10 m around the access tube, which corresponds to the range of the electromagnetic field (DELTA-T, 2008). As the access tubes are positioned 0.40 m from the stem of the plants, it is possible that at a closer distance to the stem, higher moisture contents will be recorded.

Guimarães et al. (2010) and Costa (2014) warn that measurements made by moisture sensors using the principle of reflectometry can be affected by variations in the density and composition of the soil, rocks near the stems, roots, air pockets, waste in the subsoil, and to a lesser extent, influenced by the ionic conductivity of the salts dissolved in the soil moisture, when it exceeds 250 mS m^{-1} . However, the analysis of the concentration of soluble salts in the soil's saturation extract, at the depths of 0.0 to

0.2 and 0.2 to 0.4 m, showed values below 0.20 mS m^{-1} , characterising the soil as non-saline, as discussed by Bernardo et al. (2013) and Mantovani et al. (2013), therefore not interfering with the soil moisture reading.

All the plants subjected to the treatments with water from the cistern and without water were mulched with coconut shells, covering a radius of 1.5 metres from the stem of the plant,

create a microclimate and prevent moisture loss. Therefore, it is believed that without this procedure, lower humidity levels would have been recorded in all periods and species cultivated.

In addition to the evapotranspiration process, which is boosted above all by solar radiation, surface temperature, relative air temperature and humidity and wind (Santos & Carlesso, 1998), the decrease in moisture content in the superficial part of the soil profile, observed in all periods, may also be associated with soil structure. Its lower bulk density, as well as its greater total porosity, contribute to an increase in hydraulic conductivity, rapidly draining water into the underlying layers, also driven by the greater quantity of coarse soil fractions.

In addition to the factors already mentioned, the low humidity of the soil is also related to the rationed volume of water allocated to the crops. Given the limited volume of 52,000 litres of water in the P1+2 cistern, it is not possible to meet the evapotranspiration demand of the crops, which is commonly practised in conventional irrigated agriculture.

During the experiment, which included the rainy, intermediate and dry periods, soil moisture values very close to the Permanent Wilting Point - PMP - were observed at some point. This concept presupposes the minimum limit of water availability for plants, when the water tension in the soil is approximately 1,500 kPa. However, studies by Sykes (1969) and Griffin et al. (1989) apud Procópio et al. (2004), emphasise that different plant species can develop strategies to extract water at lower limits, such as Nicotiana attenuata, Cassia fasciculata, and Agropyron intermedium, which managed to extract water from the soil at 1.610 kPa, 3,470 kPa and 3,860 kPa, respectively, through modifications to their morphological and physiological characteristics, which determine their competitive abilities for soil water.

It is possible that the plants that did not receive water from the cistern during the intermediate and dry periods resisted and managed to adapt to the low humidity conditions recorded at 0.40 metres from the stem. It is also worth noting that soil moisture values, when close to the PMP, were mainly found at 0.10 metres in the soil profile for most crops. Also in this context, research carried out by Van Lier & Libardi (1997) and Procópio et al. (2004) shows that the potential of water in the soil at PMP can vary significantly between soils, species, phenological stage of the plant, history of water stress and plant adaptation to the environment.

In general, the increase in soil moisture up to a depth of around 0.30 m, where the highest soil moisture values were observed throughout the experiment, may have been due to the number of smaller diameter pores in the 0.20 to 0.40 m layer (Table 5). Compared to the top layer, this layer has a higher silt fraction and twice as much clay fraction, which, because they have larger specific surfaces and more pores, increase the capillarity effect and increase the water retention capacity (Libardi, 2005).

Under the conditions studied, the volume of water applied from the cistern, plus rainfall, does not meet the plants' needs. However, the aim is to make it possible to maintain fruit trees on a small scale, with a minimum water requirement, with a quantity sized based on the volume of water stored in the P1+2 cistern (52,000 litres), in order to promote the inclusion of fruit and vegetables in the diet of rural families in the Brazilian semi-arid region.

4.3.2 Vegetable beds

Water was supplied to the vegetables using a plastic watering can, as shown in Figure 11. 32 litres of water were applied per day to bed 1 and 50% of this volume to bed 2, distributed in equal parts in the morning and afternoon, according to **Error! Reference source not found**. Water application totalled 15,840 litres throughout the experiment.

Figure 11 - Partial view of the vegetable beds

Photo: Nilton de Brito Cavalcanti (2014).

In order to infer the relationship between the volumes of water in each bed and the behaviour of the soil's volumetric humidity, we used data from the Soil Water Retention Characteristic Curve (CCRAS) (Graph 3) for the 0.0 to 0.2 m layer, considering that for the vegetables under study, the effective depths of the root systems are within this depth interval (Pires et al., 1999). The data used was Field Capacity - FC of 0.2565 cm^3 cm^{-3} and Permanent Wilting Point - PWS 0.0772 cm^3 cm^{-3} , with the lower limit of the soil water availability factor f = 0.5 considering a scenario of high water demand, which corresponds to 65% of Field Capacity. This is the minimum humidity for the vegetables not to have their yields jeopardised.

Although the volume of water administered to bed 2 was 16 litres of water per day, half the volume applied to bed 1, Graph 10 shows that there was no significant difference in the soil moisture values recorded between 0.10 and 0.20 m depth. The averages for these values were 0.2262 cm^3 cm^{-3} for bed 1 and 0.2087 cm^3 cm^{-3} for bed 2. As the beds were raised approximately 0.10 m from the ground, it is possible that this condition led to rapid percolation of water in this space, showing similar moisture values in depth for each bed.

From a depth of 0.20 m for bed 2, there was a noticeable loss of soil moisture as a result of the lower volume of water applied compared to that applied to bed 1, associated with water consumption by the plants and evapotranspiration processes (Santos & Carlesso, 1998). Even though the volume of water applied - a total of 5,280 litres throughout the experiment - was half that applied to bed 1, the moisture levels in the soil profile remained within the minimum limit of water availability up to 0.40 m, providing good development without affecting production. It is possible to adopt the water application criterion for bed 2, as in addition to meeting the water demands for the vegetable group, it also has soil moisture beyond the effective root system depth zone adopted for the vegetables.

For bed 1, the volume applied of 32 litres of water per day totalled a consumption of 10,560 litres throughout the experiment, and was responsible for maintaining a higher level of humidity at greater depth, showing low oscillation up to 0.40 m, ranging from 0.2262 cm^3 cm^{-3} to 0.2104 cm^3 cm^{-3} , followed by a sharp decrease in soil humidity.

It is possible that the decrease in soil water content is associated with water consumption by the plants and, above all, with the loss of soil moisture due to evaporative processes, thus helping to redistribute lower levels of moisture to the lower layers (Santos & Carlesso, 1998). From a depth of 0.60 m, the moisture levels in beds 1 and 2 showed the same behaviour, reaching a depth of 1.0 m with an average soil moisture of 0.1356 cm^3 cm^{-3} , corresponding to around 50% of Field Capacity.

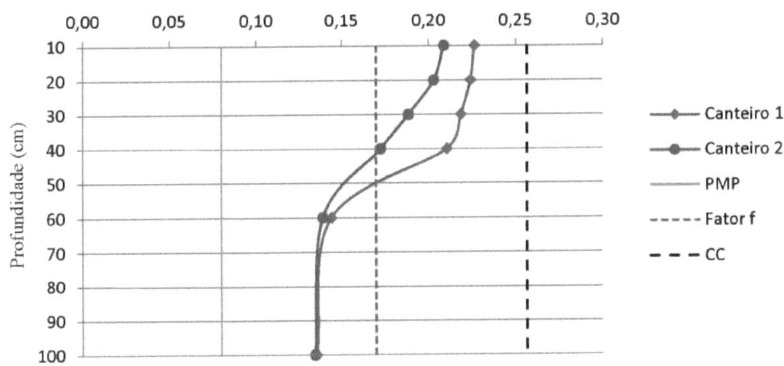

Graph 10 - Soil moisture profile in the vegetable garden, in beds 1 and 2

4.4 Vegetative development of the orchard

Below is data on the vegetative development of all the plants in the orchard, those that received water from the cistern and those that didn't, referring to one year of observation, when the experiments were conducted with the management of the water stored in the cistern. Table 7 shows the data at the start of the experiment and Table 8 shows the data at the end of the study.

Table 7 - Orchard vegetative development data at the start of the experiment

Cultures	Pl.	Stem: Height (m)	Stem: Circumference (cm)	Stem: Diameter (cm)	Cup: Height (m)	Cup: Diameter (cm)
	1	2,11	28,32	12,03	1,71	3,1
	2	2,83	29,18	12,05	2,31	3,23
Mango	3	2,88	27,42	10,08	2,79	3,48
	4	2,91	34,12	12,03	2,7	3,51
	5	2,34	27,56	10,02	2,25	4,41
Pine cones	1	2,78	38,39	13,09	2,59	2,41
	2	3,77	30,12	11,97	2,85	2,11
	3	2,41	29,45	10,91	2,11	2,25
	4	2,38	31,49	11,21	2,77	3,1
	5	2,12	24,78	10,05	2,71	3,71
	1	1,86	35,71	12,87	1,54	3,01
	2	2,35	34,18	11,05	2,05	2,91
Cashew	3	2,11	36,79	11,03	1,95	3,21

	4	2,32	30,15	11,43	1,96	2,97
	5	1,97	38,35	10,07	1,92	3,12
Acerola	1	1,32	25,16	9,01	1,25	1,9
	2	1,53	24,38	10,03	1,41	2,81
	3	1,61	18,45	6,34	1,55	2,11
	4	1,47	24,12	7,51	1,45	2,05
	5	1,55	21,17	9,09	1,41	2,11
Orange	1	0,96	9,17	2,35	0,62	1,12
	2	0,88	11,46	3,11	0,67	1,33
	3	0,91	15,78	3,89	0,77	1,48
	4	0,84	7,08	6,46	0,63	0,94
	5	0,58	7,76	4,27	0,54	0,62

Pl. = plant; Alt. = height in metres; Circumf. = circumference in cm; Diâm. = diameter in cm.

Table 8 - Orchard vegetative development data at the end of the experiment period

		--------------Stem----------------			---------Cup----------	
Cultures	**Pl.**	Height (m)	Circumference (cm)	Diameter (cm)	Height (m)	Diameter (cm)
Mango	1 a	2,64	43,04	15,01	2,22	3,70
	2 a	2,99	37,07	13,05	2,39	3,60
	3 a	3,58	34,02	12,08	2,68	3,75
	4 b	3,12	37,00	13,00	2,68	3,3
	5 b	*	*	*	*	*
Pine cones	1 a	2,93	45,21	14,90	2,70	3,56
	2 a	3,90	38,04	14,04	3,08	3,60
	3 a	2,78	38,01	12,07	2,61	3,45
	4 b	2,53	35,02	12,11	2,98	3,12
	5 b	2,96	29,04	11,03	2,76	3,82
Cashew	1 a	2,12	46,15	15,12	1,69	4,60
	2 a	2,73	42,08	15,24	2,46	4,45
	3 a	2,60	41,01	14,13	2,37	4,40
	4 b	2,45	35,02	12,60	2,33	4,30
	5 b	2,18	43,00	12,00	2,21	4,17
Acerola	1 a	1,75	32,14	13,15	1,68	3,97
	2 a	2,00	34,21	13,98	1,94	4,10
	3 a	1,96	27,09	12,04	1,98	3,80
	4 b	1,89	27,03	12,04	1,87	3,25
	5 b	1,91	27,10	12,59	1,84	3,17
	1 a	1,10	13,09	12,01	0,98	1,47

Orange	2 a	1,12	15,12	15,01	0,97	1,89
	3 a	1,78	19,18	19,11	1,68	1,98
	4 b	1,10	10,00	10,00	0,95	1,12
	5 b	0,68	9,120	7,10	0,78	0,73

Pl. = plant; Alt. = height in metres; Circumf. = circumference in cm; Diâm. = diameter in cm; (a) = plants that received water from the cistern; (b) = plants that did not receive water from the cistern; (*) = plant that did not resist the lack of water.

By comparing the data collected at the beginning and end of the experiment (Table 7 and Table 8), it can be seen that the plants of the five crops (fruit trees) that received water from the cistern showed better individual variation in stem and crown growth than the plants that didn't receive water. The best results were seen for plants 1a, 2a and 3a of the acerola, cashew and pine crops, which showed a higher individual growth rate than the other crops, even for those 4b and 5b that did not receive water from the cistern, especially in terms of stem circumference, along with crown height and diameter.

It is possible that this particularity is associated with the peculiar resistance of these crops in the field when faced with low water content in the soil. Data presented by Teixeira & Azevedo (1995) and Guanziroli et al. (2009) show that these species are resistant to periods of severe water deficiency and have an important ability to extract water from the soil.

This has meant that, in addition to better vegetative development than other crops, these crops have maintained significant production, as will be discussed below.

Of the plants that didn't receive water from the cistern, the only one that didn't resist the water shortage was mango plant 5b. The other plants of the same crop (1a, 2a, 3a and 4b), although they resisted, showed low stem and crown growth, with the exception of 4b, which reduced its crown, due to the plant's limitation in maintaining or producing new leaf area in the face of low soil humidity. This slowdown in development was also recorded for the orange plants (4b and 5b) in the treatment that did not receive water from the cistern, a behaviour that influenced the lower production performance of these crops, as will be discussed below.

4.5 Fruit and vegetable production

4.5.1 Growing fruit trees

By supplying water from the P1+2 cistern and applying it to the fruit trees incrementally over the three periods of the year, at 8, 12 and 16 litres per plant, three times a week, it was possible to obtain production for all the species that received water from the cistern (Table 9). The application of water in the intermediate and dry periods with 12 and 16 litres per plant, respectively, led to low humidity values in the soil profile, up to a depth of approximately 0.40 m, in the order of 60% below Field Capacity.

Over the course of the experiment, it was possible to keep the plants in production, obtaining a total of 262.73 kg of fruit (Table 9), especially acerola, which contains a high content of ascorbic acid (Vitamin C). This reality of fruit production can be included in the menu of families of small farmers in the Brazilian semi-arid region, despite the occurrence of a major drought in previous years.

Table 9 - Production data for three plants that received water from the cistern

Cultures	Plants	No. of fruits	Average weight (Kg)	Total weight (Kg)
	1	2	0,15839	0,317
Mango	2	17	0,15231	2,590
	3	5	0,15687	0,785
			Subtotal:	3,692
	1	110	0,17652	19,418
Pine cones	2	138	0,17831	24,607
	3	109	0,17942	19,557
			Subtotal:	63,582
	1	107	0,18951	20,278
Cashew	2	59	0,18742	11,058
	3	20	0,18763	3,753
			Subtotal:	35,089
	1	7.626	0,00778	59,331
Acerola	2	6.762	0,00754	50,989
	3	6.548	0,00684	44,789
			Subtotal:	155,109
	1	2	0,17730	0,355

Orange	2	11	0,17251	1,898
	3	18	0,16670	3,000
			Subtotal:	5,253
Total (Kg)				262,73

In view of the production obtained, it can be seen that the acerola tree (Malphigia glabra L.), the pine tree (Annona squamosa) and the cashew tree (Anacardium occidentale) showed better vegetative development, culminating in a greater quantity of fruit per unit of production, reaching a production of 155.109 kg; 63.582 kg and 35.089 kg, respectively, and to a lesser extent the orange tree (Citrus sinensis) 5.253 kg and the mango tree (Mangifera indica L.) 3.692 kg (Table 9).

The peculiar characteristics of the acerole tree (Malphigia glabra L.), the pine tree (Annona squamosa) and the cashew tree (Anacardium occidentale), in terms of their resistance and adaptability to the region's climate conditions, combined with the addition of rationed water from the production cistern, have allowed them to remain productive at appreciable levels. The other species, although also commonly cultivated by rural families in the region, and included in the research, did not respond well to the volumes of water administered during the year.

However, the presence of a production cistern in P1+2 is an alternative that can improve the diet of rural families through the daily consumption of fruit, considering that the total fruit production obtained from three plants was 262.73 kg during the year, which corresponds to the possibility of consuming 0.72 kg of fruit per day per family.

Although the plants received a water supply throughout the year, according to the sizing based on the volume of the P1+2 cistern (**Error! Reference source not found.**), it should be emphasised that the production levels obtained here, although considered low when compared to the conventional irrigation system, are important. They reflect the extreme conditions of soil moisture and other inputs needed to increase production potential to which the plants were subjected.

Figure 12 shows some of the orchard plants with fruit: (a) mango tree; (b) pine tree; (c) orange tree; (d) cashew tree and (e) acerola tree.

Figure 12 - Production aspects of the orchard

Source: Elvis Pantaleão Ferreira (2014); Nilton de Brito Cavalcanti (2014).

Table 10 shows the production data for the treatments that did not receive water from the cistern, as they only received rainfall. It can be inferred that even with the low levels of moisture recorded in the soil profile, especially during the intermediate period and especially during the dry period, there was a not insignificant production for the acerola (Malphigia glabra L.) and pine (Annona squamosa) crops.

Table 10 - Production data for non-irrigated fruit tree treatments

Cultures	Plants	No. of fruits	Average weight (Kg)	Total weight (Kg)
Mango	1	-	-	-
	2*	-	-	-
				Subtotal-
	1	89	0,17530	15,602
Pine cones	2	75	0,17221	12,916
			Subtotal	28, 518
Cashew	1	5	0,16792	0,839
	2	-	-	-
			Subtotal	0, 840
	1	3.682	0,00629	23,160
Acerola	2	4.267	0,00637	27,181
			Subtotal	50,341
	1	-	-	-
Orange	2	-	-	-
			Subtotal	-
Total				79,70

* The plant was unable to withstand the low levels of moisture in the soil.

For the cashew tree (Anacardium occidentale), only one plant produced. The other species did not produce. For the mango tree (Mangifera indica L.), one plant was unable to withstand the low levels of humidity in the soil. According to Bernardo et al. (2013) one of the most important factors observed in low crop yields is directly correlated with water supply. Although no studies have been carried out on the spatial distribution of the root system of the cultivated species, and other morphological and physiological characteristics, it is assumed that for the acerole and pine tree crops there is a differential capable of determining competitive abilities to extract water from the soil, because even subjected to the same conditions of low soil humidity as the other crops, they achieved significant production.As a result of not applying water from the cistern to these plants, when comparing the individual yields for the acerola (Malphigia glabra L.) and pine (Annona squamosa) crops, it can be seen that they produced around 40 per cent and 50 per cent less than the plants that received water from the cistern. This reinforces the fact that cisterns can make a significant contribution to food production for consumption by rural families in the semi-arid region.

4.5.2 Growing vegetables

The cultivation and production data for the vegetables grown in beds 1 and 2 with rainwater stored in the cistern are shown in Table 11. For the vegetable species whose unit of production measurement was the bunch, also known as a bundle, these were standardised at around 40 mm in diameter. Although with uniform diameters, the green weight varied according to each species: rocket 200 g/mould; leaf cabbage 180 g/mould; lettuce 225 g/mould; coriander 190 g/mould and the average weight of the chilli fruit was 95 g.

Table 11 - Production of vegetables grown with irrigation

Vegetable species	Production obtained				
	Unit	Construction site 1 (Un.)	Construction site 2 (Un.)	Total (Un.)	Total (Kg)
Arugula (Eruca sativa)	Sauce	56	42	98	19,6
Leaf cabbage (Brassica oleracea)	Sauce	48	54	102	18,36
Chilli (Capsicum annuum)	Fruit	94	43	137	13,01
Coriander (Coriandrum sativum)	Sauce	173	102	275	52,25
Lettuce *(Lactuca sativa)*	Molho	42	34	76	17,10
Vegetable species	**General information on cultivation**				

	Type of planting	Spacing (cm x cm)	Home harvest (days)[1]	Productio n Kg/m *2	Producti on Kg/m^2 **
Arugula (Eruca sativa)	Seedling s	25 x 25	80	2,8	2,1
Leaf cabbage (Brassica oleracea)	Seedling s	30 x 30	90	2,43	2,16
Chilli (Capsicum annuum)	Seedling s	25 x 25	100	2,24	1,02
Coriander (Coriandrum sativum)	Direct/south	10/Line	50	8,22	4,84
Lettuce *(Lactuca sativa)*	Mudas	25 x 25	90	2,37	1,91

[1] After germination or transplanting; *Bed 1; **Bed 2.

In general, all the vegetable species grown showed good growth. Among them, coriander showed the highest yield, reaching 8.22 Kg/m^2 for bed 1 and 4.84 Kg/m^2 for bed 2, followed by rocket and lettuce, which showed 2.80 Kg/m^2 and 2.37 Kg/m^2 for bed 1, and 2.10 Kg/m^2 and 1.91 Kg/m^2 for bed 2, respectively. In view of the yields obtained, it is possible to suggest that one of the reasons why families in the region usually prefer to grow coriander, as seen during the field visits, is associated with its higher yield and shorter time to harvest, and the consequent possibility of selling the surplus, which also improves family income.

The yields obtained from beds 1 and 2 totalled 120.32 kg during the experiment, as shown in Table 11. Of this total, 59.08% corresponded to production in bed 1 and 40.93% to production in bed 2. There was therefore no significant difference between the yields. It was observed that the volume of water dimensioned and applied in the form of 32 litres per day for bed 1 and 50% of this volume for bed 2, had little influence on the production of the vegetables, showing a production increase of around 20%, considered low, for a volume of 50% more water applied.

Therefore, the recommendation to apply only one layer of water corresponding to 4 mm/day, as dimensioned for bed 2, subdivided into two instalments, can be adopted by the families covered by P1+2, especially considering that this was a year of extreme drought.

During the experiment, 10,560 litres of water were used to grow the vegetables in bed 1 and 5,280 litres in bed 2, giving a total consumption of 15,840 litres of water throughout the experiment. Considering the individual vegetable production of each bed, we have a ratio of 148.54 litres of water per kilo of vegetables produced for bed 1, and 107.24 litres of water per kilo of vegetables produced for bed 2.

Therefore, considering the water/production ratio, bed 2 showed the best results. It is important to know the relationship between water consumption and vegetable production because the vast majority of vegetables have a low capacity to extract water from the soil, requiring more frequent water application (Marouelli et al., 2006; Donato, 2014). On the other hand, most of the water applied is lost through evapotranspiration and infiltration into the soil.

According to Silveira et al. (2011), Brazil had an average vegetable consumption of 27.08 kg/per capita/year in 2010. In the Northeast, annual per capita consumption was 22.08 kg, which corresponds to 60.49 grams per day per inhabitant. Considering the production obtained for bed 2 alone, with 49.23 kg of vegetables, it is possible to supply a family of four with a daily consumption of 33.71 grams, which is a little more than double the supply when compared to the annual per capita average of vegetables for the Northeast region. This promotes the consumption of healthy foods that are important for food and nutritional security by rural families in the Brazilian semi-arid region.

In a scenario where only bed 2 was cultivated, consuming only 5,280 litres, a balance of 10,560 litres of water in the cistern would be obtained, which, when resized to be applied to the 15 fruit trees in the orchard, could increase the supply of water applied during the dry season from 16 litres per plant to 24.3 litres, for 20 weeks, at a frequency of 3 times a week. With this significant increase in the supply of water to be applied to the fruit trees, it is believed that in addition to improving the availability of water in the soil, it would also lead to an increase in fruit production.

Figure 13 shows some images of the beds cultivated with the different vegetable species: (a) coriander; (b) cabbage and rocket; (c) peppers and lettuce; (d) peppers and (e) lettuce, (f) leaf cabbage.

Figure 13 - Production aspects of beds (1) and (2)

Source: Nilton de Brito Cavalcanti (2014).

Based on the analysis of the production data obtained in the research, cultivating just one 4 m bed[2] , with alternating plantings of rocket (Eruca sativa), leaf cabbage (Brassica oleracea), chilli (Capsicum annuum), coriander (Coriandrum sativum), lettuce (Lactuca sativa), or with others, if the farmer wishes, as long as 16 litres of water are applied per day, divided into two times, morning and afternoon, for 365 days totalling a volume of 5.840 litres, it is possible for a cistern with a capacity of 52,000 litres of water, as is currently being built by the P1+2 Programme in different communities in Brazil's semi-arid region, to provide a diversified crop of healthy fruit and vegetables. This would provide significant and regular consumption, which is important for promoting and maintaining health.

4.6Contextualisation of the use of water from the P1+2 cistern by communities in the microregion of the Vale do Submédio São Francisco

The fruit trees commonly grown in the communities are those offered to families after the construction of the production cistern has been completed, such as the acerole tree (Malpighia emarginata), coconut tree (Cocos nucifera), orange tree (Citrus x sinensis), mandarin tree (Citrus reticulata), pine tree (Annona squamosa), papaya tree (C. papaya), cashew (Anacardium occidentale), mango (Mangifera indica L.) and passion fruit (Passiflora edulis f.flavicarpa), as well as others introduced by the families. At the time of this research, three to four specimens of each species were observed.

An episode of lack of planning for the proper use of water from the cistern marked the Lindolpho Silva community in 2012. In the expectation that the water from the cistern would provide production for commercialisation, some families planted passion fruit (`Passiflora edulis f. flavicarpa`) extensively, with up to 100 passion fruit seedlings per family, as reported by Brito et al. (2012). This number of plants resulted in a lack of water in the cistern, which is necessary to complete the crop cycle. This forced some families to resort to buying water via a water tanker. This practice became economically unsustainable with the increase in water demanded by the plants during their development, resulting in the total loss of plantations in some families' orchards.

Data published by Brito et al. (2010) show that the water stored in a production cistern, if well managed, is enough to maintain a small orchard with approximately 30 fruit trees. However, according to Brito et al. (2012) an orchard with a maximum of 25 plants is recommended from 2012 onwards, considering that it is preferable to grow a small orchard than to lose most of the fruit in years of atypical rainfall, a common occurrence in the region, similar to 2012, when rainfall totalled only 26.86% of the historical average.

After the crisis and low production of the passion fruit crop, a new option for producing food and increasing the families' income was used from June 2013 onwards. They began a timid, semi-confined breeding of free-range chickens in consortium with the productive backyards, which soon became a growing activity, given the birds' peculiar adaptation to the region's climate. Allied to this was the guarantee that the meat would be sold through the federal government's Food Acquisition Programme (PAA), as well as the good reception of the eggs on the local market and consumption by the families.

The PAA is a programme coordinated by the MDS's National Secretariat for Food and Nutritional Security, established by Law 10.696/2003, which promotes the purchase of food produced by family farmers directly or through their associations and/or cooperatives, eliminating the presence of middlemen who often established a predatory and unfair relationship with producers. The products are sold at a rewarding price and used to build up government stocks or for social programmes (CONAB, 2010).

According to the National Supply Company - Conab (2010), the PAA is an instrument that has been strengthening local and regional segments of family farmers, as well as agrarian reform settlers, foresters, aquaculturists, extractivists, artisanal fishermen, indigenous people, members of rural quilombo communities and other traditional peoples and communities.

Data collected from community president Lindolpho Silva indicates that there are a total of 1,200 poultry in the community, including broilers and layers, raised in consortium with the productive backyards. The poultry is commercialised between 90 and 120 days (Figure 14) at local fairs and with the PAA, guaranteeing the purchase of 5,000 reais per family per year. According to the data collected, the quick return on sales has been motivating poultry farming in the community, unlike traditional goat farming, which takes between 1.5 and 2 years.

Figure 14 - Fruit trees in consortium with chicken farming in the Lindolpho Silva community

Photo: Elvis Pantaleão Ferreira (2014).

It was also reported that the success experienced by the families is attributed to the training courses organised in the community by trade unions, SENAI, PRORURAL, ATER, EMBRAPA and NGOs, which, through the P1+2 production cistern, guarantee water for the poultry's consumption, which rises as the rate of growth of poultry production increases. However, it is important that poultry farming is limited to the water carrying capacity of the cistern, as there is also a need to have water available for crop use.

It is worth noting that the training offered to the community enabled them to learn about poultry farming, including health management, preparing feed on the property, keeping records and notes, and using poultry manure to fertilise crops. In this way, the families have been gaining experience, skills and ability to improve their poultry farming. It was noted that this community has an entrepreneurial profile with a collective structure, which is a great competitive advantage.

Data released by the Regional Institute of Small Appropriate Agriculture - IRPAA (2012), which works with small family farmers in the semi-arid Sertão do São Francisco, shows that since 2012 various projects have been implemented in the region to raise free-range chickens. According to the Bahia Agricultural Development Company (EBDA) (2014), chicken farming is an important segment of the economy. provide families with a protein-rich diet, and their excrement is used as fertiliser for crops. In addition, this activity adds income to family farmers through the commercialisation of their products.

The production cisterns present in the communities have also been helping to maintain goat and sheep farming during periods of prolonged drought, presenting themselves as a complementary alternative to family income. However, despite the progress, the research showed that some of the P1+2 Programme cisterns in the communities studied are unproductive.

As far as the Campo Verde community is concerned, family testimonies about the cultivation of fruit trees revealed problems with termite attacks on plants, causing damage to crops and leading to the death of some. According to Ferreira & Barrigossi (2006), some species of termite are considered to be soil pests, characterised as a group of insects capable of causing damage to the underground part of the plant, roots and stems, which can lead to a severe reduction in production and even death of the plant.

These same authors report that attacks by rhizophilous termites (Syntermes molestus) are a major concern given the threats they pose to plant maintenance, especially in sandy soil and low humidity conditions, which makes it easier for them to move around the area. These characteristics can be observed in the locality, but there is no exact information on termite infestation in the region, as the data is limited to family information. This fact was not emphasised in the other communities.

According to the families, termite attacks on fruit plants have discouraged cultivation, because although they have used agro-ecological techniques, such as the use of chickens, natural dressings with garlic and onions, among others, they have not been able to control them and have opted to grow vegetables, the surplus production of which has in some cases been commercialised. Testimonies and data collected in the communities revealed that although there are fruit plants in the backyards, cultivated with water from the cistern, the Baixa da Boa Vista and Campo Verde communities were more adept at growing vegetables. Specifically in the cultivation of spring onions (Allium cepa), coriander (Coriandrum sativum), lettuce (Lactuca sativa), peppers (Capsicum annuum) and sporadically, tomatoes (Solanum lycopersicum), as they pointed out that it was difficult to grow the latter.

It should be noted that it is also common to plant medicinal herbs for use by the family, such as rosemary (Rosmarinus officinalis), mallow (Malva sylvestris), pomegranate (Punica granatum), among others, irrigated with water from the cistern. Some families used coconut straw and/or shade cloth to soften the sun on the vegetable beds (Figure 15).

Figure 15 - Cultivation of coriander and peppers in the Baixa da Boa Vista (A) and Campo Verde (B) communities

Photo: Elvis Pantaleão Ferreira (2014).

Reports from families in these communities emphasise that until mid-2012, growing vegetables made it possible to meet family consumption, share with relatives and sell surplus produce at local fairs. They reported that in previous years there had been a certain regularity of rainfall, allowing for a larger area to be cultivated and for the cistern to be filled more frequently. However, there were reports that "*from 2012 to the present day, the rains have been scarce*", jeopardising the recharging of the cisterns and the water supply.

The water from the cisterns was used to grow vegetables only for the family's consumption. However, despite the low rainfall in the region, it is possible to grow permanent crops on a smaller scale with the water from the cistern.

Data published by the Ministry of Agriculture, Livestock and Supply - MAPA (2013) corroborates that in 2012 and 2013 rainfall in the semi-arid region was well below the historical average, which was experienced until 2014, the year of this research, jeopardising the practice of agricultural activities, especially those of family farmers. As a result, it disrupted the rural economy and affected the recharge of the reservoirs that supply rural and urban communities. These facts were recorded in the interviews with the communities.

Testimonies from the communities revealed that the P1+2 cistern was a great help in the face of prolonged droughts and the lack of water in traditional water sources, which are generally not perennial, allowing livestock to be kept and food to be produced. As recorded in the testimony of a resident of Campo Verde "the *cistern was a great blessing, because without it there would be nothing to harvest in the backyard"* and another resident of the Baixa da Boa Vista community, who said *"the cistern made it possible for many families to stay in the countryside*".

The research also revealed that some P1+2 cisterns are not being properly used to grow fruit and vegetables or to water animals, but are being used as a second source of water for domestic purposes. The data collected indicates that the increase in the drilling of artesian wells and the construction of new dams may be leading some families to use the water from the production cistern for domestic purposes.

In general, all the communities visited do not use pesticides on their crops, but only adopt agroecological practices in their farming activities, such as mulching (Figure 16) and fertilising with poultry and small ruminant manure. As well as promoting better plant nutrition, reducing soil moisture loss through evaporation and increasing the conservation of natural resources, these practices contribute to a better quality of life through the production and consumption of healthy food.

Figure 16 - Mulch used on fruit trees

Photo: Elvis Pantaleão Ferreira (2014).

the P1+2 production cisterns have helped families stay in the countryside and contributed to their food and nutritional security, since they guarantee regular and permanent access to healthy food, such as fruit and vegetables, free from agrochemicals. In this sense, studies carried out by Araújo et al. (2011) and Brito et al. (2012) confirm the possibility of growing various species of fruit and vegetables on a small scale, making significant improvements to families' diets and helping to reduce possible illnesses, especially among children, whose diet needs to be diversified in terms of vitamins and minerals.

As for how water is supplied to crops, there is no clear methodology. The form of application varies greatly in frequency and volume, with crops being irrigated with buckets, watering cans and hoses. In this sense, it was observed that in the communities present in the three municipalities, although the water is applied using buckets and watering cans, usually every other day, many families draw water from the water source.

cistern using a$^{1}/_{2}$ CV (Steam Horse) water pump, powered by electricity, directing the water to the plants using a hose (Figure 17).

Figure 17 - Using a pump to draw water from cisterns

Photo: Elvis Pantaleão Ferreira (2014).

It's worth pointing out that the practicality of using a pump with a hose has contributed to an excessive amount of water being spent on the plants, as there is no mechanism for controlling the volume applied. Added to this are some cases where many plants are grown and the water from the cistern is used for domestic purposes.

In this context, it's worth noting that a simple and easy-to-use technique used at

Embrapa Semiárido's Caatinga Experimental Field to help growers with the timing and volume of water applied to plants is the use of a plastic container with a defined volume to be applied. This strategy prevents the producer from using water impulsively. Because of the above, there are testimonies that the 52,000 litres of water in the cistern is not enough to grow fruit, vegetables and maintain small herds of goats and sheep. Because of this, it is necessary for some families to refill the cistern with water tankers, when there are financial resources available. However, in the P1+2 programme's conception, water is to be used for one activity or another, and that the use of water is sustainable.

Data presented by El-Deir (2012) and Gheyi (2012) show that the rural environment for small family farmers in the semi-arid region has undergone technological changes in recent years, with the introduction of inputs and irrigation systems, driven mainly by rural credit aimed at modernising production methods and improving quality of life. It is therefore necessary to introduce technologies based on training and monitoring farmers in the correct use of these techniques.

Although all the families receiving production cisterns have been trained in training courses and exchanges in the proper use of the cisterns, there is a need for better technical pedagogical support. In this way, reflective links will be established to guide and discipline families so that the water is administered and sized correctly for crops and livestock, so that the water resource does not run out during the dry season. According to Alvarenga et al. (2011), dialogue with reality, at times when knowledge is socialised using simple language, is an important tool for reflecting on the possibilities for change and innovation to improve the solidity of any programme.

Experiences reported by El-Deir (2012) and Gheyi et al. (2012) show that the process of transforming rural communities in the semi-arid region can take place through extension activities and the structuring of collective projects whose responsibility is shared by all. They also emphasise that the work of non-governmental organisations is decisive, considering their commitment to the region and the strengthening of relationships of trust with the community, acting as facilitators of the process of collective construction in informal spaces.

Finally, it can be seen that the implementation of the P1+2 cisterns has been an important instrument for keeping the population in the countryside and improving the families' diet, as well as strengthening the bonds of identity for living in the region. However, for the programme to be more successful and solid, it is essential to systematically monitor the families so that they can make effective and appropriate use of the cisterns, as well as a methodology for using the water from the cisterns to produce food.

CHAPTER 5

CONCLUSIONS

The water content recorded in the soil profile for the irrigated treatment during the periods considered was highest during the rainy season, with a value of 0.2584 cm^3 cm^{-3} recorded at 0.3 m depth. For the same depth in the intermediate and dry periods, 0.1741 cm^3 cm^{-3} and 0.1087 cm^3 cm^{-3} were recorded respectively.

It was possible to obtain 262.73 kilos of fruit from 15 fruit trees, with the acerola (Malphigia glabra L.), pine (Annona squamosa) and cashew (Anacardium occidentale) crops standing out as having the highest yields per unit.

In the soil profile for bed 1, where 32 litres of water per day were applied, higher soil moisture contents were recorded at greater depths, reaching 0.4 m at a value of 0.2104 cm^3 cm^{-3} . For bed 2, which received 50% of the water from bed 1, the moisture content corresponded to 0.1724 cm^3 cm^{-3} , with no significant difference between the yields.The criterion for applying water to bed 2, as well as meeting the water demands for the vegetable group, resulted in a balance of 10,560 litres of water in the cistern, which could allow the application of water to fruit trees to be resized during the dry season.

The P1+2 Programme's cistern provides perennial and diversified cultivation of healthy fruit and vegetables, enabling significant and regular consumption, which is important for promoting and maintaining the health of the rural population.

There is a need for better technical pedagogical support regarding the rational and efficient use of water from the P1+2 cistern, so that the water is administered and sized correctly for crops and livestock, so that the water resource does not run out during the dry season.

REFERENCES

IYFFA - **International Year for Family Farming**. Available at << http://www.aiaf2014.gov.br/>>. Accessed on 12 April 2015.

ALVARENGA, A. T.; VASCONCELLOS, M. P.; ADORNO, R. C. F. A contribuição das ciências sociais e humanas na pesquisa, na ensino e na formação em saúde. **Revista Saúde e Sociedade**. v. 20, p. 9 - 11, n° 1. São Paulo Jan./Mar. 2011.

AMARAL FILHO, J. ASSIS JÚNIOR, R. N.; MOTA, J. C. A. **Soil Physics**: Concepts and Applications. Fortaleza - CE. University Press, 2008. 290 p.

AMORIM, J. R. A.; CRUZ, M. A. S.; RESENDE, R. S.; BASSOI, L. H.; SILVA FILHO, J. G. Spatialisation of the percentage of exchangeable soil sodium in the California Irrigated Perimeter, in Canindé de São Francisco, Sergipe - SE. **Research Bulletin**, 61. Embrapa Tabuleiros Costeiros, p. 17, Dec. 2010.

ARAÚJO, F. S.; RODAL, M. J. N.; BARBOSA, M. R. V.; MARTINS, F. R. Distribution of woody flora in the caatinga domain. **Analysing variations in the biodiversity of the caatinga biome:** Support for regional conservation strategies. Ministry of the Environment - MMA, Brasília, 2005.

ARAÚJO, J. O.; BRITO, L. T. L.; CAVALCANTI, N. B. Rainwater stored in cisterns can increase the nutritional quality of families' diets. **Cadernos de Agroecologia**, Cruz Alta, v. 6, n. 2, p. 1 - 5, Fortaleza, dec. 2011.

ASA - **Articulação Semiárido Brasileiro**. Available at << http://www.asabrasil.org.br/Portal/Informacoes.asp?COD_MENU=1151>>. Accessed on 23 November 2014.

AUDRY, P.; SUASSUNA, J. Water quality in irrigation in the semi-arid tropics - A case study. In: Franco-Brazilian Seminar on Irrigation Research, 1990, Recife. **Proceedings of the Franco-Brazilian Seminar on Small Irrigation, Research and Development**. Recife - PE, 11 to 13 December 2009.

AYERS, R.S.; WESTCOT, D.W. Water quality in agriculture. 2.ed. Campina Grande: UFPB, FAO. **Irrigation and Drainage Studies**, 29 revised 1. 1999, 153p.

BARROS, G. A. C. Family farming. **Centre for Advanced Studies in Applied Economics** - CEPEA, ESALQ/USP. p. 1 - 3, 2006.

BARROS, M. F. C.; FONTES, M. P. F.; ALVAREZ, V. H.; RUIZ, H. A. Recovery of soils affected by salts through the application of gypsum from quarrying and limestone in north-eastern Brazil. **Brazilian Journal of Agricultural and Environmental Engineering**, v. 8, p. 59 - 64, 2006.

BERNARDO, S.; SOARES, A. A.; MANTOVANI, E. C. **Irrigation Manual**. 8th Ed. 5th reprint. Viçosa - MG: Ed. UFV, p. 625, 2013.

BERTONI, J. & LOMBARDI NETO, F. **Soil conservation.** 6.ed. São Paulo, Ícone. 2008. 355 p.

BRAZIL. **Decree No. 63.778, of 11 December 1968**. Provides for the inclusion of municipalities in the area of the Polígono das Secas. Available at <<http://www2.camara.leg.br/legin/fed/decret/1960-1969/decreto-63778- 11-December-1968-405144-publicacaooriginal-1-pe.html>>. Accessed on 12 April 2015.

BRAZIL. **Decree-Law No. 9.857, of 13 September 1946**. Provides for the creation of the National Department of Works Against Droughts. Available at <<http://www.planalto.gov.br/ccivil_03/decreto-lei/1937- 1946/Del9857.htm>>. Accessed on 12 April 2015.

BRAZIL. **Law No. 11.326, of 24 July 2006**. Establishes the guidelines for formulating the National Policy for Family Farming and Rural Family Enterprises.

BRAZIL. **Law No. 175 of 7 January 1936**. Provides for the defence plan against the effects of drought. Available at <<http://www.camara.gov.br/sileg/integras/230371.pdf>>. Accessed on 12 April 2015.

BRAZIL. Ministry of National Integration. **New delimitation of the Brazilian semi-arid** region. Brasília, DF, p. 32, 2005.

BRAZIL. **Ordinance No. 89 of 16 March 2005**. Updates the list of municipalities belonging to the Semi-Arid region of the Northeast Constitutional Financing Fund.

BRITO, L. T. L.; ARAÚJO, J. O. A.; CAVALCANTI, N. B.; SILVA, M. J. Rainwater stored in a cistern produces fruit and vegetables for consumption by rural families: A case study. **8th Brazilian Symposium on Rainwater Harvesting and Management**. Campina Grande - PB, 14 to 17 August 2012.

BRITO, L. T. L.; CAVALCANATI, N. B.; PEREIRA, L. A.; GNADLINGER, J.; SILVA, A. de SOUZA. **Rainwater stored in cisterns for fruit and vegetable production**. Petrolina - PE: Embrapa Semiárido (Documentos, 230), p. 30, 2010.

CAMPOS, J. N. B. Droughts and public policies in the semi-arid region: ideas, thinkers and periods. **Estudos Avançados**. v. 28, n. 82. 2014. 65 - 88 p.

CAMPOS, J. N. B. Vulnerabilidades Hidrológicas do Semi-Árido às Secas. **Revista Planejamento e Políticas Públicas/Instituto de Pesquisa Econômica Aplicada (Ipea).** Brasília DF - Brazil. n° 16, p.1 - 38, 1997.

CAPOBIANCO, J. P. R. The Brazilian biomes. In: CAMARGO, A.; CAPOBIANCO, J. R. P.; OLIVEIRA, J. A. P. (Org.) **Meio ambiente Brasil**: avanços e obstáculos pós-Rio-92. São Paulo: Estação Liberdade: Instituto Socioambiental/FGV. 2002. 117 - 155 p.

CARDOSO, J. A. F.; LIMA, A. M. N.; CUNHA, T. J. F; AMARAL, A. J.; OLIVEIRA NETO, M. B.; HERNANI, L. C. Organic carbon in the humified fractions of organic matter in sandy soils under mango cultivation in the Brazilian semi-arid region. **I Reunião Nordestina de Ciência do Solo.** 22 to 26 September 2013, CCA/UFPA - Areia/PB.

CAVALCANTI, F. C. (Coord). **Fertiliser recommendations for the state of Pernambuco**. 2ª Approximation. Agronomic Institute of Pernambuco - IPA. Recife, PE, p. 198, 1998. Available at << http://www.ipa.br/publicacoes_tecnicas.php>>. Accessed on 27 March 2015.

CAVALCANTI, N. B.; BRITO, L. T. L.; ARAÚJO, J. O. Production of fruit trees irrigated with rainwater in the semi-arid region of the northeast. **8th Brazilian Symposium on Rainwater Harvesting and Management**. Campina Grande - PB, 14 to 17 August 2012.

CINTRA F. L. D.; NEVES, C. S. V. Aspectos metodológicos do estudo do sistema radicular de plantas perenes através de imagens. **Boletim informativo da SBCS**, Campinas, v. 21, n° 3, 1996.

COELHO, E. F.; OLIVEIRA, F. C.; ARAUJO, E. C. E.; VASCONCELOS, L. F. L. Distribution of "Pêra" orange roots under rainfed and micro-sprinkler irrigation in sandy soil. **Pesq. agropec. bras.** v.37, n.5, 2002, 603 - 611 p.

COELHO, E. F.; SIMÕES, W. L.; CARVALHO, J. E. B.; COELHO FILHO, M. A. **Root distribution and soil water extraction in tropical fruit trees under irrigation**. Embrapa Mandioca e Fruticultura Tropical - Cruz das Almas/BA, p. 80, 2008.

CONAB - **National Supply Company.** Food Acquisition Programme - PAA. Basic Legislation. 162 p. Brasília-DF, July 2010.

COOK, F. J. Modelling trickle irrigation: comparison of analytical and numerical models for estimation of wetting front position with time. **Environmental Modelling & Software**, v. 21, n. 9, p. 1353 - 1359, 2006.

CORRÊA, R.M.; NASCIMENTO, C.W.A.; ROCHA, A.T. Adsorption of phosphorus in ten soils of the State of Pernambuco and its relationship with physical and chemical parameters. **Acta Sci. Agron**. Maringá, v. 33, 2011. 153 - 159 p.

CORREIA, R. C.; KIILL, L. H. P.; MOURA, M. S. B.; CUNHA, T. J. F.; JESUS

JUNIOR, L. A. de; ARAUJO, J. L. P. **The Brazilian semi-arid region.** In: VOLTOLINI, T. V. (Ed.). Petrolina: Embrapa Semiárido, p. 21 - 48, 2011.

COSTA, A. B, (Org.) **Social Technology and Public Policies.** São Paulo: Pólis Institute; Brasília: Banco do Brasil Foundation, 2013.

COSTA, B. R. S. Calibration of a capacitance sensor for measuring humidity in semi-arid soils. Juazeiro - BA, Dissertation (**Master's in Agricultural Engineering**) - Federal University of the São Francisco Valley, Campus Juazeiro - BA, 2014. 122 p.

CUNHA, T. J. F.; PETRERE, V. G.; SILVA, MENDES, A. M. S.; MELO, R. F.; OLIVEIRA NETO, M. B.; SILVA, M. S. L.; ALVAREZ, I. A. **Main soils of the Brazilian tropical semi-arid region:** characterisation, potential, limitations, fertility and management. In: SÁ, I. B.; SILVA, P. C. G. (Eds.). Brazilian semi-arid region. Petrolina, PE: Embrapa Semiárido, p. 49 - 87, 2010.

CUNHA, U. S.; MACHADO, S. A.; FIGUEIREDO FILHO, A. Use of exploratory data analysis and robust regression in the evaluation of the growth of commercial dryland species in the Amazon. **R. Árvore**, Viçosa-MG, v. 26, n. 4, p. 391 - 402, 2002.

DELTA - T. **User Manual for the Profile Probe type PR2.** Delta-T Devices Ltd, 2008.

DIACONIA. **Construction of a 52,000 litre cistern**. Living with the Semi-Arid. Sharing Experiences Series n° 5. Family Farming Support Programme - PAAF. Recife - PE, p. 1 - 49, 2008.

DONAGEMMA, G. K.; RUIZ, H. A.; FONTES, M. P. F.; KER, J. C.; SCHAEFER, C. E. G. R. Dispersion of Latosols in response to the use of pre-treatments in textural analysis. **Revista Brasileira de Ciência do Solo**, v. 27, p. 765 - 772, 2003.

DONATO, L. **Teaching Bulletin No. 88 - Vegetable production on the southern coast of Santa Catarina**. Agricultural Research and Rural Extension Company of Santa Catarina

Catarina - Epagri. Available at <<http://www.epagri.sc.gov.br/>> Accessed on 26 April 2015.

EBELING, A.; A. M. KLEIN; J. SCHUMACHER; W. W. WEISSER & T. TSCHARNTKE. How does plant richness affect pollinator richness and temporal stability of flower visits? **Oikos.** v. 117, 2008. 1808 - 1815 p.

EBELING, A.G.; ANJOS, L.H.C.; PEREZ, D.V.; PEREIRA, M.G. & VALLADARES, G.S. Relationship between acidity and other chemical attributes in soils with high levels of organic matter. **Bragantia**, v. 67, 2008. 261 - 266 p.

ECHART, C. L; MOLINA, S. C. Aluminium phytotoxicity: effects, tolerance mechanism and genetic control. **Ciência Rural, Santa Maria**, v.31, n.3, p. 531 - 541 2001.

EDBA - **Empresa Baiana de Desenvolvimento Agrícola**. Available at << http://www.ebda.ba.gov.br/>>. Accessed on 23 November 2014.

EL - DEIR, S. G. (Org.). **Environmental Education in the Semi-Arid:** Methodological Proposals for Rural Extension. 1st ed. Recife, EDUFRPE, p. 256, 2012.

EL - DEIR, S. G. **Innovative Methodologies for Social Empowerment**. 1st ed. Recife, EDUFRPE, p. 237, 2013.

EMBRAPA - Brazilian Agricultural Research Corporation. National Soil Research Centre. **Manual of soil analysis methods** / National Soil Research Centre. - 2. ed. rev. atual. - Rio de Janeiro, p 212, 1997.

EMBRAPA - Brazilian Agricultural Research Corporation. **Manual of Chemical Analysis of Soils, Plants and Fertilisers.** Brasília: Embrapa Solos/Embrapa Informática Agropecuária/Embrapa Comunicação para Transferência de Tecnologia, p.370, 1999.

ERTHAL, V. J. T.; FERREIRA, P. A.; MATOS, A. T. Physical and chemical alterations of an Argissolo by the application of cattle wastewater. **Rev. bras. eng. agríc. ambient.** v. 14, n. 5, p. 467 - 477, 2010.

FAO - **Food and Agriculture Organisation**. Family farming in the spotlight on World Food Day. Available at << http://www.fao.org/family-farming-2014/news/news/details-press- room/en/c/254637/>>. Accessed on 14 April 2015.

FAO - **Food and Agriculture Organisation**. The arid environments. In Arid zone forestry: A guide for Field technicians. Food and Agriculture

Organisation of the United Nations, Via delle Terme di Caracalla. Rome, Italy. Available at <<http://www.fao.org>>. Accessed on 27 March 2015.

FBB - Banco do Brasil Foundation. **Social Technology**. Available at << http://www.fbb.org.br/tecnologiasocial/o-que-e/tecnologia-social/>>. Accessed on 20 May 2015.

FERREIRA, E; BARRIGOSSI, J. A. F. **Orizivorous Insects of the Underground Part**. Documents 190. Embrapa Rice and Beans. Santo Antônio de Goiás, GO. p. 52, 2006.

FREDLUND, D. G., & XING, A. Predicting the permeability function for unsaturated soils using the soil-water characteristic curve. **Canadian Geotechnical Journal**, v. 31, n° 3, 1994. 521 - 532 p.

GERSCOVICH, D. M. S. Equations for modelling the characteristic curve applied to Brazilian soils. **IV Brazilian Symposium on Unsaturated Soils**, Porto Alegre, RS, March 2001.

GHEYI, H. R.; PAZ, V. P. S.; MEDEIROS, S. S.; GALVÃO, C. O. (Ed.) **Water Resources in Semi-Arid Regions**. Campina Grande - PB. Instituto Nacional do Semiárido - INSA, Universidade Federal do Recôncavo Baiano - UFRB. p. 249 - 268, 2012.

GIL, A. C. **Métodos e Técnicas de Pesquisa Social**. 6ª ed. São Paulo: Atlas, p. 220, 2008.

GIULIETTI, A. M.; QUEIROZ, L. P.; SANTOS, S. T. R.; FRANÇA, F.; GUEDES, M. L.; AMORIM, A. M. Flora of Bahia. **Sitientibus,** v.6, n° 3, 2006. 169 - 173 p.

GNADLINGER, J.; SILVA, A. de S.; BRITO, L. T. de L. **P1 + 2: One Land, Two Waters Programme for a sustainable semi-arid region.** In: BRITO, L. T. de L.; MOURA, M. S. B. de; GAMA, G. F. B. (Ed.). Potential of rainwater in the Brazilian semi-arid region. Petrolina: Embrapa Semi-Arido, chap. 3, p. 63 - 77, 2007.

GUANZIROLI, C. E.; SOUZA, H. M.; VALENTE JÚNIOR, A.; BASCO, C. A. Barriers to the development of cashew farming in the Northeast: commercialisation margins or increases in productivity and scale. **Revista Extensão Rural**, DEAER/PPGExR - CCR - UFSM, Ano XVI, n° 18, p. 96 - 112, Jul - Dec 2009.

GUANZIROLI, C. H. & CARDIM, S. E. de C. S. (Coord.). **New portrait of family farming:** Brazil rediscovered. INCRA/FAO Technical Co-operation Project, MDA Ministry of Agrarian Development, Brasília, DF: INCRA/FAO, MDA, 2000.

GUIMARÃES, P. L. O.; SANTANA, M. A. A.; OLIVEIRA, I. R.; THOMAZ JÚNIOR, J. C. Proposed procedure for calibrating soil moisture sensors and meters (Soil Moisture). **Metrology Quality Congress São Paulo State Metrological Network** - REMESP. 25 to 27 May 2010, São Paulo, Brazil.

HOMEM, B. G. C.; ALMEIDA NETO, O. B.; SANTIAGO, A. M. F.; SOUZA, G. H. Clay dispersion caused by fertigation with wastewater from animal farms. **Revista Brasileira de Agropecuária Sustentável** (RBAS), v.2, n. 1, p. 89 - 98, July, 2012.

HORST, W. J.; WANG, Y.; ETICHA, D. The role of the root apoplast in aluminium-induced inhibition of root elongation and in aluminium resistance of plants: **a review**. Annals of Botany 106: 2010. 185 - 197 p.

IBGE - **Brazilian Institute of Geography and Statistics.** Available at < http:

http://www.ibge.gov.br/home/> Accessed on 17 March 2014.

IBGE, **Information base of the 2010 Demographic Census**: Population in the Demographic Censuses, according to Major Regions and Federation Units.
http://www.censo2010.ibge.gov.br/sinopse/index.php?dados=4&uf=00>.
Accessed in September 2014.

INCRA - National Institute for Colonisation and Agrarian Reform. **Geographical Variation in the Size of Fiscal Modules in Brazil**. Available at < http://www.incra.gov.br/media/institucional/legislacao/atos_internos/instrucoes /instrucao_especial/IE20_280580.pdf>. Accessed in May 2015.

INSA - National Semi-Arid Institute. **Desertification and climate change in the Brazilian semi-arid region.** Campina Grande: INSA - PB, p. 209, 2011.

IPEA - Institute for Applied Economic Research. **Brazilian Agriculture:** Performance, Challenges and Prospects. GASQUES, J. G.; VIEIRA FILHO, J. E. R.; NAVARRO, Z. (Org.). Brasília: Ipea, p. 78, 2010.

IRPAA - **Instituto Regional da Pequena Agropecuária Apropriada**. Available at <<http://www.irpaa.org/>>. Accessed on 23 November 2014.

JACOMINE, P. K. T. **Soils under caatinga:** Characteristics and agricultural use. In: ALVAREZ V., V. H.; FONTES, L. E. F. & FONTES, M. P. F. (Eds.). Soil in the great morphoclimatic domains of Brazil and sustainable development. Viçosa, MG, SBCS/UFV, p. 169 - 199, 1996.

JAQUES, R. C. **Quality of rainwater in the municipality of Florianópolis and its potential for use in buildings**. Postgraduate Programme in Environmental Engineering (Master's dissertation). Federal University of Santa Catarina. Florianópolis, SC. p. 102, 2005.

KAMIMURA, A.; OLIVEIRA, A.; BURANI, G. F. A. **Family farming in Brazil** - a portrait of regional imbalance. Interações, Campo Grande, vol.11, n.2, 2010. 217 - 223 p.

KLAUS, B. Calcium in soils and plants. Research Centre Hanninghof, Yara International, Germany. **Agronomic information**, n. 117, 2007. 212 - 252 p.

KLEIN, V. A. **Física do solo**, Passo Fundo: Ed. Universidade de Passo Fundo, p. 212, 2008.

KLUTE, A. **Water retention**: laboratory methods. In: Klute, A. (ed.). Methods of soil analysis. 2.ed. Madison: American Society of Agronomy, 1986. Part. 1, 635 - 662 p.

KONRAD, M.; HERNANDEZ, F. B. T.; SANTOS, R. A. Spatial distribution of the acerola root system in a Yellow Red Podzolic soil. **XXX Brazilian Congress of Agricultural Engineering - CONBEA**. Mabu Thermas & Resort, Foz do Iguaçu - Paraná, 31 July to 03 August 2001.

LEAL, I. R.; TABARELLI, M.; SILVA, J. M. C. (Eds.). **Ecology and conservation of the Caatinga**. Recife: Ed. Universitária da UFPE, p. 822, 2003.

LEAL, I. R.; TABARELLI, M.; SILVA, J. M. C.; LACHER JR, T. E. Changing the course of biodiversity conservation in the Caatinga of Northeast Brazil. **Megadiversidade**, Belo Horizonte, v. 1, n. 1, p. 139 - 146, 2005.

LIBARDI, P. L. **Water dynamics in the soil**. São Paulo - SP. EDUSP, p. 335, 2005.

LOIOLA, M. I. B.; ROQUE, A. A.; OLIVEIRA, A. C. P. Caatinga: Vegetation of the Brazilian semi-arid region. **Revista Ecologi@** - Online Journal of the Portuguese Ecology Society. n. 4, Jan - Apr, p. 85, 2012.

LOYOLA, J. M. T.; PREVEDELLO, C. L. Analytical models for predicting the process of soil water redistribution. **Rev. Bras. Ciênc. Soil**. v. 27, n. 5, p. 783 - 787, 2006.

MACEDO, J. R. General Characteristics of the Fertility of Sandy Soils in Brazil. **Brazilian Symposium on Sandy Soils.** Embrapa Soils. President

Prudente, SP. São Paulo, 2014. Available at: http://www.unoeste.br/site/destaques/outros_destaques/documentos/PALESTRAS_SBSA_2014_PDF/07.pdf >>. Accessed on 28 March 2015.

MAIA, C. E.; LEVIEN, S. L. A. Wet bulb dimensions in surface drip irrigation. **Revista Ciência Agronómica**, v. 41, n. 1, p. 149 - 158, 2010.

MALAVOLTA, E. **Manual de nutrição mineral de plantas**. São Paulo: Ed. Agronômica Ceres, 2006. 638 p.

MANTOVANI, E. C.; BERNARDO, S.; PALARETTI, L. F. **Irrigation - Principles and methods**. 3 ed. updated - Viçosa, MG: UFV, p. 355, 2013.

MAPA - MINISTRY OF AGRICULTURE, LIVESTOCK AND SUPPLY. **Report on Drought in the Northeast - No. 37**. Secretariat for Agricultural Policy. General Coordination of Agricultural Studies and Information. Brasília - DF, 2013.

MARINHO, L. B.; MORAES, S. O. **Studies on the DIVINER 2000TM capacitive probe for determining soil moisture under deficit irrigation.** Luiz de Queiroz College of Agriculture/UPS. Piracicaba - SP. 2009. 7 p.

MAROUELLI, W. A, SILVA, W. L. C, SILVA, H. R. **Manejo da irrigação em hortaliças**. EMBRAPA-SPI/EMBRAPA-CNPH, Brasília, Brazil, 5. ed. 1996. 72 p.

MARTINS, C. M.; GALINDO, I. C. L.; SOUZA, E. R.; POROCA, H. A. Chemical and microbial attributes of soil from areas undergoing desertification in the semi-arid region of Pernambuco. **Rev. Bras. Ciênc. Solo**, Viçosa, v. 34, n. 6, p. 1883 - 1890, 2010.

MAY, T. **Social research:** issues, methods and processes. Porto Alegre: Artmed, p. 124, 2004.

MDS - **Ministry of Social Development and Fight against Hunger** (2000). Available at << http://www.mds.gov.br/falemds/perguntas-frequentes/bolsa- familia/programas-complementares/beneficiario/agricultura-familiar>>. Accessed on 18 March 2015.

MDS - Ministry of Social Development and Fight against Hunger. **Zero Hunger: A Brazilian Story**. Ministry of Social Development and Fight against Hunger, Zero Hunger Advisory. Brasília, DF: v. 2, 2010.

MEURER, E. J. Potassium. In: FERNANDES, M. S. **Nutrição Mineral de Plantas**. Brazilian Society of Soil Science, Viçosa - MG, 2006. 281 - 295 p.

MI - Ministry of National Integration. **CONVIVER**, Programme for the Integrated and Sustainable Development of the Semi-Arid Region. Secretariat for Regional Programmes - SPR. Brasília - DF, 2009. 43 p.

MI - Ministry of National Integration. **New Delimitation of the Brazilian Semi-Arid**. Secretariat for Regional Development Policies. Brasília, January 2005.

MIGUEL, M. G.; TEIXEIRA, R. S.; PADILHA, A. C. C. Suction characteristic curves of lateritic soil from the Londrina/PR region. **Revista de Ciência & Tecnologia**. v. 12, n° 24, p. 63 - 74. 2006.

MILLER, C. J.; YESILLER, N.; YALDO, K.; MERAYYAN, S. Impact of soil type and compaction conditions on soil water characteristic. **Journal of geotechnical and geoenvironmental engineering.** v. 128, n° 9, 2002.

MOLIN, J. P.; GIMENEZ, L. M.; VOLNEI, P. U. R. S.; JENS H. Measurement of soil electrical conductivity by induction and its correlation with production factors. **Eng. Agríc., Jaboticabal**, v.25, n.2, 2005. 420 - 426 p.

MOTA, F. O. B.; OLIVEIRA, J. B. Mineralogy of soils with excess sodium in the State of Ceará. **Revista Brasileira de Ciência do Solo**, vol. 23, n° 4, 1999, 799 - 806 p.

MOURA, M. S. B.; GALVINCIO, J. D.; BRITO, L. T. L.; SOUZA, L. S. B. DE.; SÁ, I. I. S.; SILVA, T. G. F. **Climate and rainwater in the Semi-Arid**. In: BRITO, L. T.

de L.; MOURA, M. S. B. de.; GAMA, G. F. B. (Ed.). Potential of rainwater in the Brazilian Semi-Arid. Petrolina: Embrapa Semi-Arido, chap. 2, p. 37 - 59, 2007.

MUALEM, Y. A new model for predicting the hydraulic conductivity of unsaturated porous media. **Water Res.** v. 12, p. 513 - 522, 1976.

NASCIMENTO, T.; BRITO, L. T. L.; CAVALCANTI, N, B. Manejo de Água na Irrigação de salvação nas unidades do P1+2. **X Congreso Latinoamericano y del Caribe de Ingeniería Agrícola** and XLI Congresso Brasileiro de Engenharia Agrícola CLIA/CONBEA 2012 Londrina - PR, Brazil, 15 to 19 July 2012.

OLALDE, A. R. **Family farming and sustainable development.** 2011. Available at << http://www.ceplac.gov.br/radar/Artigos/artigo3.htm>>. Accessed on 23 January 2015.

OLIVEIRA, A. C.; FÁRIAS FILHO, S. M. A parallel between the fruit producers of the Petrolina Juazeiro cluster and the fruit growers of the entire BNB area. **Journal of Economic Development.** Department of Applied Social Sciences - Year XIV, No. 26, December 2012. Salvador, BA.

OLIVEIRA, F. C.; MATTIAZZO, M. E.; MARCIANO, C. R.; ROSSETTO, R. Effects of successive applications of sewage sludge on a dystrophic Yellow Latosol cultivated with sugarcane: organic carbon, electrical conductivity, pH and CTC. R**. Bras. Ci. Solo**, v. 12, n. 24, p. 63 - 74. 2002.

PAN BRASIL. **National action programme to combat desertification and mitigate the effects of drought.** Brasília, DF: Ministry of the Environment, Secretary of Water Resources, p. 242, 2004.

PEREIRA, J. R. **Soils affected by salts**. In: Cavalcanti, F.J.A. (Coord). Fertiliser recommendations for the state of Pernambuco: 2ª approach.2 ed. Recife: IPA, p. 76 - 82, 1998.

PIRES, R. C. M.; SAKAI, E.; ARRUDA, F. B.; FUJIWARA, M.; CALHEIROS, R. O. **Methods and management of irrigation**. Centre for Ecophysiology and Biophysics. Agronomic Institute/IAC. p. 32, 1999.

PIRES, R.C.M.; CALHEIROS, R. O.; SAKAI, E.; FUJIWARA, M. & ARRUDA, F.B. **Basic information on irrigation and drainage**. In: Instruções Agrícolas para o Estado de São Paulo, 6 ed. Campinas, Instituto Agronômico/IAC. p. 222 - 225, 1999.

PRADO, D. **The caatingas of South America**. In: LEAL, I. R.; TABARELLI & SILVA, M. J. M. C. (eds.). Ecology and conservation of the Caatinga. Recife: Ed. Universitária da UFPE, p. 139 - 146, 2003.

PRADO, R. M.; CENTURION, J. F. **Changes in the colour and degree of flocculation of a Dark Red Latosol under continuous sugarcane cultivation**. Pesquisa Agropecuária Brasileira, v. 36, p. 197 - 203, 2001.

PREZOTTI, L. C.; GOMES, J. A.; DADALTO, G. G.; OLIVEIRA, J. A. **Manual de recomendação de calagem e adubação para o estado do Espírito Santo - 5° Aproximação**. Vitória - ES, p. 305, 2007.

PREZOTTI, L. C.; GUARÇONI, M. A. **Guia de interpretação de análise de solo e foliar**. Vitória - ES, Incaper, p. 104, 2013.

PROCÓPIO, S. O.; SANTOS, J. B.; SILVA, A. A.; DONAGEMMA, G. K.; MENDONÇA, E. S. Permanent Wilting Point of Soya, Beans and Weeds. **Planta Daninha,** Viçosa - MG, v. 22, n. 1, p. 35 - 41, 2004.

REICHARDT, K. & TIMM, L. C. **Soil, plant and atmosphere:** Concepts, processes and applications. São Paulo, Manole, p. 478, 2012.

RHEINHEIMER, D. S.; & SOUZA, R. O. Electrical conductivity and acidification of waters used for herbicide application in Rio Grande do Sul. **Ciência Rural**, Santa Maria, v. 30, n. 1, p. 97 - 104, 2000.

RHEINHEIMER, D.S. & ANGHINONI, I. Distribution of inorganic phosphorus in soil management systems. **Pesq. Agropec. Bras.**, v. 36, 2001. 151 - 160 p.

SALVADOR, J. T.; CARVALHO, T. C.; LUCCHESI, L. A. C. Calcium and magnesium relations in soil and macronutrient leaf contents. **Rev. Acad. Agrár. Ambient**. Curitiba, v. 9, n. 1, p. 27 - 32, jan./mar. 2011.

SANTOS, A.; GOIS, F. F. **Microcrédito e Desenvolvimento Regionais**: Instituto para o Desenvolvimento de Estudos Econômicos, Sociais e Políticas Públicas. Fortaleza: Premius 1, p, 384, 2011.

SANTOS, D. B.; COELHO, E. F.; AZEVEDO, C. A. V. Water absorption by lemon tree roots under different irrigation frequencies. **Rev. bras. eng. agríc. ambient.**, v. 9, n.3, 2005. 327 - 333 p.

SANTOS, H. G. dos.; JACOMINE, P. K. T.; ANJOS, L. H. C. dos.; OLIVEIRA, V. A. de; OLIVEIRA, J. B.; COELHO, M. R.; LUMBRERAS, J. F.; CUNHA, T. J. F. (Ed.). **Brazilian soil classification system**. 2.ed. Rio de Janeiro: Embrapa Solos, 2006.

SANTOS, R. F.; CARLESSO, R. Water deficit and the morphological and physiological processes of plants. **Revista Brasileira de Engenharia Agrícola e Ambiental**, v.2, n.3, p.287 - 294, 1998. Campina Grande, PB, DEAg/UFPB.

SILVA, A. de S.; MOURA, M. S. B.; BRITO, L. T. de L. **Lifesaving irrigation for**

subsistence crops. Petrolina - PE: Embrapa Semiárido, Chap. 8, 2006. 159 - 179 p.

SILVA, C. R. da; ANDRADE JÚNIOR, A.S. de; SOUZA, C. F. **Practical aspects of using the capacitance technique:** challenges and learning. In: II Workshop on Applications of Electromagnetic Techniques for Environmental Monitoring. Taubaté-SP, 2008. CD-ROM.

SILVA, D. F.; GALVÍNCIO, J. D.; ALMEIDA, H. R. R. C. Variability of water quality in the São Francisco river basin and related anthropic activities. **Qualit@s**. Electronic Journal, v. 9, n. 3, 2011.

SILVA, R. M. A. **Between combat and coexistence with the Semi-Arid: Transitions, paradigms and the sustainability of development**. Doctoral thesis. Centre for Sustainable Development. University of Brasília - Unb. Brasília, 2006.

SILVEIRA, J; GALESKAS, H; TAPETTI, R; LOURENCINI, I. **Who is the Brazilian fruit and vegetable consumer? Hortifruti Brasil**. Cepea - Centre for Advanced Studies in Applied Economics - ESALQ/USP. July 2011. Available at <http://www.cepea.esalq.usp.br/>. Accessed on 12 March 2015.

SOARES, J. M. , COSTA, F. F.; SANTOS, C. R. Manejo de irrigação em frutiras. In: Irrigation management. **XXVII Brazilian Congress of Agricultural Engineering - CONBEA**. Poços de Calda, 1998. 281 - 311 p.

SOUZA FILHO, P. S; HELDWEIN, A. B; ZAMBERLAN, J. F; CORRÊA, H. C. Soil physical parameters related to the advance of the wetting front. **R. Bras. Eng. Agríc. Ambiental**, v.17, n. 2. 2013. 153 - 161 p.

SOUZA, S. S.; TOMASELLA, J.; GRACIA, M. G.; AMORIM, M. C.; MENEZES, P. C. P.; PINTO, C. A. M. O Programa de monitoramento climático em tempo real na área de atuação da SUDENE. **Revista Brasileira de Meteorologia**, v. 25 n. 1, 2001. 15 - 24 p.

SOUZA, V. V; DIAS, H. C. T; COSTA, A. A. Analysing the quality of water from open and effective precipitation in a secondary Atlantic Forest fragment in the municipality of Viçosa, MG. **R. Árvore**, Viçosa - MG, v. 31, n. 4, 2007. 737 - 743 p.

SUDENE. **Superintendence for the Development of the Northeast**. Characterisation of the Semi-Arid Region. Available at <<http://www.sudene.gov.br/semiarido>>. Accessed on 15 March 2015.

TEIXEIRA, A. H. C. **Agrometeorological Information from the Petrolina, PE/Juazeiro, BA Hub - 1963 to 2009**. Embrapa Semiárido - Petrolina, PE. (Documentos, 233). 2010. 21 p.

TEIXEIRA, A. H.; AZEVEDO, P. V. Climate limit indices for acerola cultivation.

Pesquisa Agropecuária Brasileira, Brasília, v. 30, dec. 1995.

URACH, F. **Estimation of water retention for irrigation purposes**. Dissertation (Master's in Agricultural Engineering) - Postgraduate Course in Agricultural Engineering, Federal University of Santa Maria, RS. 2007. 78 p.

URACH, F. **Estimation of water retention for irrigation purposes**. Dissertation (Master's degree in agricultural engineering) - Postgraduate course in Agricultural Engineering, Federal University of Santa Maria, RS. 2007. 81 p.

VAN GENUCHTEN, M. T. A closed form equation for predicting the hydraulic conductivity of unsaturated soils. **Soil Science Society of America Journal**, Madison, v. 44, 1980.

VAN LIER, J. Q. & LIBARDI, P.L. Extraction of soil water by plants: Development and validation of a model. **R. Bras. Ci. Solo**, v. 21, 1997. 535 - 542 p.

VITTI, G. C.; LIMA, E.; CICARONE, F. Calcium, Magnesium and Sulphur. In: FERNANDES, M. S., ed. **Nutrição mineral de plantas**. Viçosa, Brazilian Society of Soil Science, 2006. 355 - 374 p.

VUOLO, J. H. **Fundamentals of Error Theory.** 2ª ed.: Edgar Blucher, 1996. 126 p.

WANDERLEY, N. Historical roots of the Brazilian peasantry. In: TEDESCO (Org.) Agricultura familiar: realidades e perspectivas. Passo Fundo - RS: UPF, 2001, 405 p.

WHITE, R. E. Principles and Practice of Soil Science, 4th edition. **Wiley-Blackwell**. 2005. 384 - 391 p.

Printed by Books on Demand GmbH, Norderstedt / Germany